SpringerBriefs in Molecular Science

For further volumes:
http://www.springer.com/series/8898

Wan Azlina Ahmad · Wan Yunus Wan Ahmad
Zainul Akmar Zakaria · Nur Zulaikha Yusof

Application of Bacterial Pigments as Colorant

The Malaysian Perspective

 Springer

Wan Azlina Ahmad
Department of Chemistry
Faculty of Science
Universiti Teknologi Malaysia
UTM Skudai
81310 Skudai Johor
Malaysia
e-mail: azlina@kimia.fs.utm.my

Wan Yunus Wan Ahmad
Faculty of Applied Sciences
Textile Technology Programme
Universiti Teknologi MARA
Shah Alam
Malaysia
e-mail: wanyunus@salam.uitm.edu.my

Zainul Akmar Zakaria
Institute of Bioproduct Development
Universiti Teknologi Malaysia
UTM Skudai
81310 Skudai Johor
Malaysia
e-mail: zainul@ibd.utm.my

Nur Zulaikha Yusof
Department of Chemistry
Faculty of Science
Universiti Teknologi Malaysia
UTM Skudai
81310 Skudai Johor
Malaysia
e-mail: owenikha_06@yahoo.com

ISSN 2191-5407
ISBN 978-3-642-24519-0
DOI 10.1007/978-3-642-24520-6
Springer Heidelberg Dordrecht London New York

e-ISSN 2191-5415
e-ISBN 978-3-642-24520-6

Library of Congress Control Number: 2011938995

Cover design: eStudio Calamar, Berlin/Figueres

Printed on acid-free paper

Springer is part of Springer Science+Business Media (www.springer.com)

Preface

Strong consumer demand for natural products has prompted many researchers to look for alternatives to synthetic pigments which are widely used. Synthetic pigments are not only undesirable or harmful, but can cause adverse effects to the environment. There are many sources of natural pigments, namely from micro-organisms and plants. Among plants, the orange/yellow colour obtained from saffron is one of the most expensive natural pigments sold with a price range between USD 1,100 and 11,000 per kilogram. The ascomycetous fungi, *Monascus* on the other hand has been reported to produce a variety of red, yellow, orange, green and blue hues which are mainly used in food industries.

This brief serves as a quick guide on the isolation, characterization and applications of pigments extracted from red, yellow and violet bacteria namely *Serratia marcescens*, *Chryseobacterium* sp. and *Chromobacterium violaceum*, respectively. The unique feature of this brief is the use of cheap agricultural waste for the propagation of the bacteria. This will help reduce the cost of the pigments when taken to a larger production scale. Another point to note here is the short life cycle of the bacteria which makes it viable to be used on a large scale. This brief serves as an introductory series for many more briefs on this subject.

We would like to thank each and every one for the kind help in making this publication possible especially researchers in the Universiti Teknologi Malaysia's *ColorBac* Research team namely NurZulaikha Yusof, Nordiana Nordin, Quek Hsiao Pei, Chua Pei Yong, Nurhayati Ramlee, Wong Yong Foo, Akram Neshati and NurNazrina Ahmad Sabri. A special note of thanks to Nordiana Nordin for the brilliant effort in securing funding for this coloured business. Also to the Ministry of Agriculture, Malaysia for funding of project through the Technofund grant,

TF0310F080 and Cradle Fund Sdn. Bhd. for the CIP150 grant. Not forgetting the researchers at the Program of Textile Technology, Faculty of Applied Sciences, Universiti Teknologi Mara (UiTM), Shah Alam, Selangor, Malaysia.

Johor Bahru and Shah Alam, Malaysia   Wan Azlina Ahmad
August 2011   Wan Yunus Wan Ahmad
                                                          Zainul Akmar Zakaria
                                                            Nur Zulaikha Yusof

Contents

Chapter 1
Introduction

Abstract Before the turn of the nineteenth century, natural dyes were the only source of color available and were widely used and traded, providing a major source of wealth creation around the globe. Since the introduction of synthetic dyes, many convenient and cheap synthetic pigments have appeared with Azo dyes as the most frequently used synthetic dyes on the industrial scale. However, environmental concerns regarding synthetic dyes saw a revival in the demand for natural dyes which exhibit better biodegradability and generally have a higher compatibility with the environment. Lately, the potential of obtaining natural color from microbial pigments to be used as natural colorants is being actively investigated. However, most of the bacterial pigment production is still at the R&D stage. Hence, work on the production of pigments from bacteria should be intensified especially in finding cheap and suitable growth medium which can reduce the cost and increase its applicability for industrial production.

Keywords Pigment · Synthetic · Biological · Bacteria · Dye · Natural

1.1 Definition of Pigments

The word pigment has a Latin origin and initially denoted a colored matter but it was later extended to indicate colored objects such as makeup. In the beginning of the middle ages, the word was also used to describe the diverse plant and vegetable extracts, especially those used as food colorants. The word pigment is still used in this sense in the biological terminology such as the colored matter present in animals or plants, occurring in the granules inside the cells as deposits on tissues or suspended in body fluids (Ullmann 1985). It also includes organic compounds isolated from cells and their modified structure (Hendry and Houghton 1997).

W. A. Ahmad et al., *Application of Bacterial Pigments as Colorant*,
SpringerBriefs in Molecular Science, DOI: 10.1007/978-3-642-24520-6_1,
© The Author(s) 2012

Examples of biological pigments include chlorophyll and hemoglobin. The modern meaning associated to the word pigment has its origin in the twentieth century, meaning a substance constituting of small particles which is practically insoluble in the applied medium, and is used due to its colorant, protective or magnetic properties (Ullmann 1985). Pigment also changes the color of light it reflects as a result of selective color absorption. This definition applies well to the pigments of mineral origin, such as titanium dioxide or carbon black, but for the soluble dyestuffs, usually the organic compounds, the expressions *dye, colorant* or simply *color* (as in the food colors) is more adequate. The terms pigment and color are usually applied for the food coloring matters, sometimes indistinctly (Timberlake and Henry 1986).

1.2 Classification of Pigments and Its Applications

Pigments are classified as either organic/inorganic or natural/synthetic (Turner 1993). *Biological pigments* can be classified based on two general classifications namely structural affinities and natural occurrence. Some examples for naturally occurring pigments are anthocyanins (blue–red), carotene (yellow–red), chlorophylls (green) and tannins (brown–red) (Babitha et al. 2004) (Fig. 1.1). Carotenoids are tetraterpenoids which are synthesized in plants and other photosynthetic organisms as well as in some non-photosynthetic bacteria, yeast and molds. The red and yellow appearance of autumn foliage can be attributed to the exposure of anthocyanins, as a result from the decomposition of green chlorophyll pigments, hence the removal of the masking effect.

All biological pigments selectively absorb certain wavelengths of light while reflecting others. Absorbed light may be used by the plant to power chemical reactions, while the reflected wavelengths of light determine the color the pigment will appear to the eye. Pigments also serve to attract pollinators. Carotenoids come in the forms of red, orange or yellow (tetraterpenoids). They function as accessory pigments in plants, helping to fuel photosynthesis by gathering wavelengths of light not readily absorbed by chlorophyll. The most familiar carotenoids are carotene (an orange pigment found in carrots), lutein (a yellow pigment found in fruits and vegetables) and lycopene (the red pigment responsible for the color of tomatoes) (Richella et al. 2002). Carotenoids have been shown to act as antioxidants and to promote healthy eyesight in humans. Selected carotenoids are components of the light harvesting system in chloroplasts and play an important role in the protection of plants against photooxidative damage (Demming-Adams 2002). Lycopene is able to prevent activity against several pathologies, such as cardiovascular disease, hepatic fibrogenesis and some cancer types such as prostate, gastrointestinal and epithelial (Clinton 1998). Chlorophyll is the primary pigment in plants; it is a porphyrin that absorbs yellow and blue wavelengths of light while reflecting green. It is the presence and relative abundance of chlorophyll that gives plants their green color. All land plants and green algae possess two forms of

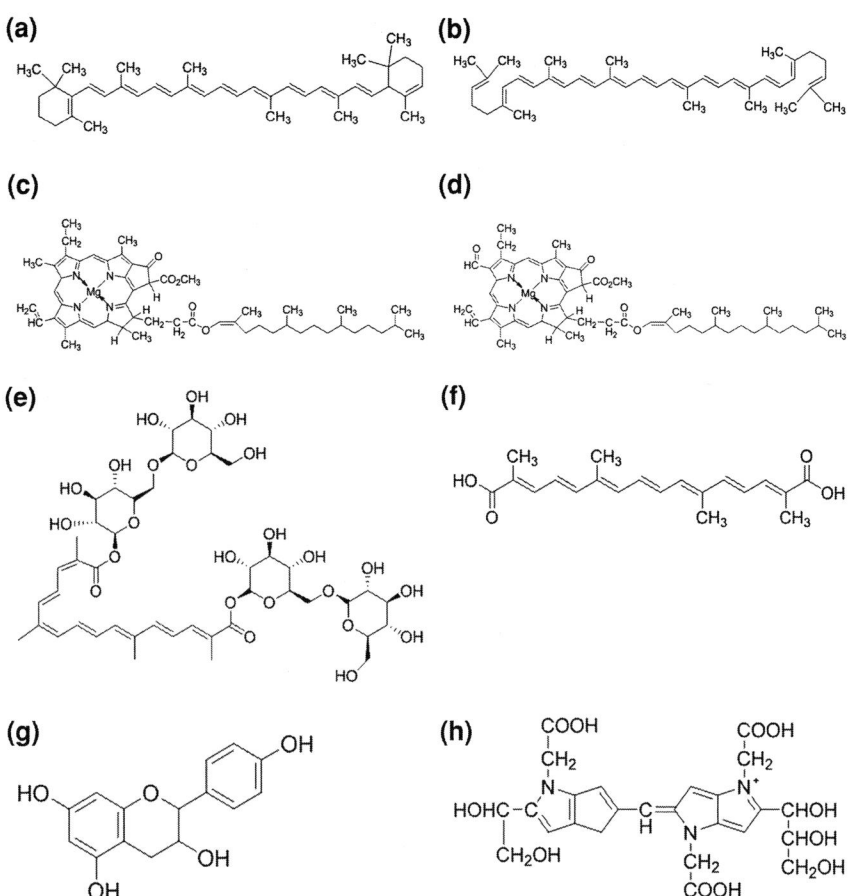

Fig. 1.1 Chemical structures for **a** ß-carotene **b** lycopene **c** chlorophyll-*a* **d** chlophyll-*b* **e** crocin **f** crocetin **g** anthocyanin **h** melanoidin

this pigment namely chlorophyll *a* and chlorophyll *b*. Kelps, diatoms and other photosynthetic heterokonts contain chlorophyll *c* instead of *b*, while red algae possess only chlorophyll *a*. All chlorophylls serve as the primary means for plants to intercept light in view of photosynthesis and also the sole reason for the green appearance of most plants (Goodwin 2002). Anthocyanins are water-soluble flavonoid pigments that have a red to blue appearance (depending on pH). It occurs in all tissues of higher plants, providing color to leaves, stems, roots, flowers and fruits, though not always in sufficient quantities to be noticeable. Anthocyanins are most visible in the petals of flowers where it constitutes as high as 30% of the dry weight of the tissue. They are also responsible for the purple color seen on the underside of tropical shade plants such as *Tradescantia zebrina*; in these plants, the anthocyanin catches light that has passed through the leaf and reflects it back towards regions bearing chlorophyll, in order to maximize the use of available light (Goodwin 2002). Betalains are indole-derived compounds synthesized from tyrosine. These water-soluble red or yellow pigments (like anthocyanins) are found only in the *Caryophyllales* (including cactus and amaranth) and never co-occur in plants with anthocyanins. Betalains are responsible for the deep red color of beets, and are used commercially as food coloring agents (Daniel 1986). Saffron, known also as CI natural yellow 6, *safran*, crocin, crocetin and crous, is the dried stigma of *Crocus sativus*, a plant indigenous to the orient but also widely grown in North Africa, Spain, Iran and France. It is a reddish, brown or golden yellow odoriferous powder with a slightly bitter taste. The stigmas of approximately 165,000 blossoms are required to make 1 kg of colorant (Daniel 1986). The coloring principles of saffron are crocin and crocetin (Fig. 1.1). Crocin is a yellow–orange glycoside that is freely soluble in hot water, slightly soluble in absolute alcohol, glycerine and propylene glycol and insoluble in vegetable oils. Crocetin is a dicarboxylic acid that forms brick-red rhombs from acetic anhydride which decomposes at about 285 °C. It is very sparingly soluble in water and most organic solvents (Daniel 1986). As a food colorant, saffron shows good overall performance. In general, it is stable towards light, oxidation, microbiological attack and changes in pH. Its tinctorial strength is relatively high, resulting in commercial application levels of 1–260 ppm (Daniel 1986). Melanoidins are ingested via consumption of brown processed food. Nutritional and physiological effects of melanoidins have been widely investigated. When rats were fed with nondialyzable melanoidins prepared from D-glucose and glycine, the difficult-to-excrete melanoidins were partly transformed into metabolizable compound (Hiromichi and Fumitaka 2002). When D-xylose and glycine were reacted at low temperature (2–26.5 °C), the reaction mixture produced yellow, red and blue pigment. Blue melanoidin has two pyrrolopyrrole rings coupled with methane bridge. The UV–vis spectrum of blue melanoidin shows a large peak at 625 nm and a small peak at 283, 322 and 365 nm.

Basically, most of the natural pigments have several features to distinguish them from the larger number of colorless compounds found in biological materials. Almost all biological molecules are composed of not more than 17 elements within the periodic table. Of these, only 4 out of 17 elements are predominant notably H, C, N and O. Most pigmented compounds (other than yellow) contain either N or O, often both and most of them are relatively large molecules with molecular weights (M_w) of 200 (anthraquinones), 300 (anthocyanidins), 400 (betalaines), 500 (carotenoids) and 800 for chlorophylls (Hendry and Houghton 1996). *Biological pigments* can be classified into six major structural classes namely tetrapyrroles, tetraterpenoids, quinines, *O*-heterocyclic, *N*-heterocyclic and metallo-proteins (Table 1.1).

Natural inorganic pigments, derived mainly from mineral sources, have been used as colorants since prehistoric times and a few, notably iron oxides, remain of some significance today. The color of inorganic pigments arises from electronic transitions which are quite diverse in nature and different from those responsible

Table 1.1 Classes of biological pigments (Hendry and Houghton 1997)

Group	Alternative or familiar name	Major example	Predominant color
Tetrapyrroles	Porphyrins and porphyrin derivatives	Chlorophylls	Green
		Heams (hemes)	Red
		Bilins (Bile pigments)	Blue–green–yellow–red
Tetraterpenoids	Carotenoids	Carotenes	Yellow–red
		Xanthophylls	Yellow
O-heterocyclic compounds	Flavonoids	Anthocyanins	Blue–red
		Flavonols	Yellow–white
Quinines	Phenolic compounds	Naphtaquinones	Red–blue–green
		Anthraquinones	Red–purple
N-heterocyclic compounds	Indigoids and indole derivatives	Betalaines	Yellow–red
		Eumelanins	Black–brown
	Substituted pyrimidines	Pterins	White–yellow
		Purines	Opaque white
Metalloproteins	Cu-proteins	Blue–green	Blue–green
	Hemerythrin	Red	Red
Miscellaneous		Lipofuscins	Brown–grey
		Fungal pigments	Various but commonly yellow

for organic pigments. Inorganic pigments generally exhibit high inherent opacity, a property which may be attributed to the high refractive index which results from the compact atomic arrangement in their crystal structures. Various methods are employed in the manufacturing of inorganic pigments. Frequently, the chemistry is carried out in aqueous solution from which the pigments can precipitate directly in a suitable physical form. In some cases, high temperature solid state reactions are used, while gas-phase processes are employed because of their suitability for continuous large-scale manufacturing. Some examples for natural inorganic pigments include titanium dioxide, carbon black, cadmium sulfides, lead chromates and ultramarines (Christie 2001).

The *synthetic organic pigments* were prepared from water-soluble dyes rendered insoluble by precipitation into colorless inorganic substrates such as alumina and barium sulfate where these products were referred to as 'lakes'. A critical event in the development of the organic pigment industry was the discovery of copper phthalocyanine in 1928. Inorganic pigments generally provide higher intensity and brightness compared to organic pigments which are unable to provide the degree of opacity, typical of inorganic pigments, because of the lower refractive index associated with organic crystals. The range of commercial organic pigments exhibit variable fastness properties depending on the molecular structure and the nature of intermolecular association in the solid state. Since organic molecules have the natural tendency to dissolve in organic solvents, organic pigment molecules are normally incorporated with structural features to enhance solvent resistance. Examples for synthetic organic pigments include azo pigments, copper phthalocyanines and high-performance organic pigments (Christie 2001).

1.3 World Scenario on the Use of Natural Pigments

Before the turn of the nineteenth century, natural dyes were the only source of color available and were widely used and traded, providing a major source of wealth creation around the globe. It has been used for many purposes such as the coloring of natural fibers (wool, cotton, silk), fur and leather. The dyes were also used to color cosmetic products and to produce inks, watercolors and artist's paints (Cristea and Vilarem 2006). Since the introduction of synthetic dyes by Perkin in 1856, many convenient and cheap synthetic pigments have appeared, and the use of natural dyes has decreased due to the relatively cheaper synthetic pigments (Zollinger 1991). Current applications of synthetic dyes are in the textile industry, leather tanning industry, paper production, food technology, agricultural research, light-harvesting arrays, photoelectrochemical cells and in hair colorings. Moreover, synthetic dyes have been employed to control the efficacy of sewage and wastewater treatment and for the determination of specific surface area of activated sludge for groundwater tracing (Forgacs et al. 2004).

Azo dyes are the most frequently used synthetic dyes on the industrial scale along with anthraquinone, sulfur, indigoid, triphenylmethyl (trityl) and phthalocyanine derivatives (Forgacs et al. 2004). Some features of the synthetic dyes include high stability to light, oxygen, pH, color uniformity, low microbiological contamination and relatively lower production costs (Alves et al. 2008). In the pharmaceutical industry, synthetic dyes were added into many medicinal products to give color as well as making it more attractive, easier to recognize, and in some cases, by forming an opaque layer, it stabilizes the ingredients of the medicine which are light sensitive (Jaworska et al. 2005). The application of synthetic dyes in the food industry is a cheaper alternative than the use of natural dyes, however, many of these dyes gives rise to serious reservations concerning health. Dyes such as tartrazine (E 102), cochineal red (E 124) and sunset yellow (E 110) may provoke allergic or pseudo-allergic reactions (PARs), either on its own or in combination with other colorants, particularly in people allergic to aspirin and other non-steroidal anti-inflammatory agents, or those suffering from urticaria or asthma (Rowe and Rowe 1994). Although, all of the synthetic colorants approved by the Food and Drug Administration (FDA) for use in foods, pharmaceuticals and cosmetic preparations have been scrutinized for its toxicity, examination on cancer chemopreventive effect of synthetic colorants revealed that a number of these products were evaluated for their in vitro antitumor promoting effect on Epstein–Barr virus (EBV) antigen induced by tumor promoter 12-O-tetradecanoylphorbol-13-acetate (TPA). Among these were the azo colorant, tartrazine (FD&C Yellow # 5) and the indigo derivative, indigo carmine (FD&C Blue # 2) (Kapadia et al. 1998). Some synthetic dyes have even been withdrawn from external usage due to its apparent hazards. For example, benzidine dyes may cause bowel cancer while carbon black (widely used as printing ink pigment) is thought to be a potential carcinogen.

From the environmental point of view, unethical discharging of untreated industrial effluents containing toxic, carcinogenic and non-biodegradable synthetic dyes into the water system is a major source of water pollution (Chung et al. 1992). One example of such dyes are the azo dyes which are widely used in industries such as textile mills (major usage), food, pharmaceutical, paper and printing, leather and cosmetics. Azo dyes are capable to retain their color and structural integrity under exposure to sunlight, soil and bacteria (resistance to microbial degradation in wastewater treatment systems), hence its classification is as recalcitrant pollutant (Eichlerová et al. 2006).

In view of this, much of the R&D efforts in wastewater treatment technologies were focused towards finding an effective and economical solution to treat the highly diverse composition of azo dye-containing industrial effluents. In spite of the existence of a variety of chemical and physical treatment processes, its removal from the environment is still extremely difficult. Adsorption, precipitation, chemical degradation or photodegradation are financially and methodologically demanding as well as time-consuming and most importantly, not very effective. Current biological approach such as the use of anaerobic bacteria which have the ability to decolorize several azo dyes may not be suitable as the bacterial reduction process of the dyes would lead to the generation of aromatic amines that are carcinogenic and mostly more toxic than the starting azo dyes (Hu 2001; Wong and Yuen 1996).

1.4 Natural Pigments

Environmental concerns regarding synthetic dyes saw a revival in the demand for natural dyes as natural dyes exhibit better biodegradability and generally have a higher compatibility with the environment (Kamel et al. 2005). Lately, the potential of obtaining natural color from microbial pigments to be used as natural colorants is being actively investigated (Nagia and EL-Mohamedy 2007). However, most of the bacterial pigment production is still at the R&D stage. Hence, work on the production of pigments from bacteria should be intensified especially in finding cheap and suitable growth medium which can reduce the cost and applicable for industrial production. Table 1.2 shows the status for microbial production of pigments:

1.5 Microbial Pigments

1.5.1 Bacteria

Several intensely colored compounds have been isolated from certain bacteria that resemble pigments in other biological systems. One example of pigment from bacteria is violacein, 3-[1,2-dihydro-5-(5-hidroxy-1H-indol-3-yl)-2-oxo-3Hpyrrol-3-ilydene]-1,3-dihydro-2H-indol-2-one, a violet pigment produced

Table 1.2 Microbial production of pigments (already in use or highly potential to be used as natural food colorants) (Liu and Nizet 2009)

Microorganism	Pigment	Color	Status
Bacteria			
Agrobacterium aurantiacum	Astaxhantin	Pink–red	RP
Paracoccus carotinifaciens	Astaxhantin	Pink–red	RP
Bradyrhizobium sp.	Canthaxhantin	Dark-red	RP
Streptomyces echinoruber	Rubrolone	Red	DS
Flavobacterium sp.	Zeaxanthin	Yellow	DS
Paracoccus zeaxanthinifaciens	Zeaxanthin	Yellow	RP
Fungus			
Monascus sp.	Ankaflavin	Yellow	IP
Monascus sp.	Monascorubramin	Red	IP
Penicillium oxalicum	Anthraquinone	Red	IP
Blakeslea trispora	Lycopene	Red	DS
Fusarium sporotrichioides	Lycopene	Red	RP
Cordyceps unilateralis	Naphtoquinone	Deep blood-red	RP
Ashbya gossypi	Riboflavin	Yellow	IP
Monascus sp.	Rubropunctatin	Orange	IP
Blakeslea trispora	ß-carotene	Yellow–orange	IP
Fusarium sporotrichioides	ß-carotene	Yellow–orange	RP
Mucor circinelloides	ß-carotene	Yellow–orange	DS
Neurospora crassa	ß-carotene	Yellow–orange	RP
Phycomyces blakesleeanus	ß-carotene	Yellow–orange	RP
Penicillium purpurogenum	Unknown	Red	DS
Yeast			
Saccharomyces neoformans var. *nigricans*	Black	Melanin	RP
Xanthophyllomyces dendrorhous	Astaxanthin	Pink–red	DS
Rhodotorula sp.	Torularhodin	Orange–red	DS

IP Industrial production, *DS* development stage , *RP* research project

Fig. 1.2 Chemical structure of violacein

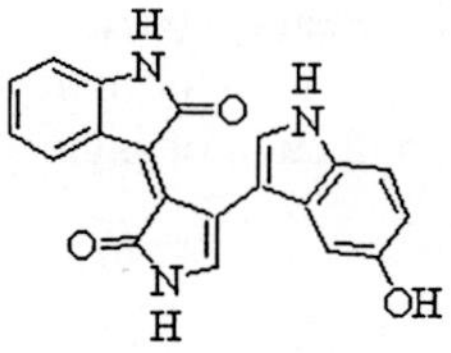

by *Chromobacterium violaceum* under aerobic condition (Duran and Menck 2001) (Fig. 1.2).

Chromobacterium violaceum is a Gram-negative proteobacteria found in the soil and water in tropical and subtropical environments. The bacterium is able to live under anaerobic and aerobic conditions but violacein production only occur in aerobic condition. Violacein is a secondary metabolite and has great potential for pharmacological applications (Buxbaum 1993).

The biosynthesis and biological properties of violacein have been extensively studied, in particular, its antitumoral, antibacterial, antiulcerogenic, antileishmanial and antiviral activities (Leon et al. 2001; Melo et al. 2003). The pigment appears to be similar to other known species of *Chromobacterium* and assisted in the identification of genus of the causative organisms. Another example of a well-studied bacterial pigment is carotenoid which imparts the eponymous golden color to the major human pathogen, *Staphylococcus aureus*. This organism produces multiple carotenoid pigments via a well described biosynthetic pathway from two tryptophan molecules. The C-3 site of each tryptophan is coupled to form a pyrrole ring where subsequent decarboxylation and 1, 2-shift of the indole ring afford the violacein chromophore (Rattanaphani et al. 2007) which culminates with golden staphyloxanthin as the major product and yellow $4'4'$-diaponeurosporene as a minor product (Wieland et al. 1994; Pelz et al. 2005). Staphyloxanthin consists of a C30 polyene carbon backbone with alternating single and double bonds typical of carotenoid pigments; these alternating bonds are able to absorb excess energy from reactive oxygen species, ROS. Violacein has one hydroxyl group and three N–H groups, which are believed to play important roles in antioxidant activity (Cao et al. 2007).

Due to the vast applications of violacein, most of the researches carried out were focused on optimizing growth conditions of *C. violaceum* to achieve maximum pigment production. *C. violaceum* is known to multiply in a growth-exacting condition. Innis and Mayfield (1979) reported that upon reincubation of *C. violaceum* in NB at 20 °C (from dormant condition at 0 °C), the violacein-less bacterium started to excrete violacein again which suggest its optimum growth temperature of 20 °C. Similar observations were reported when the bacterium was grown in soil extract broth (SEB) supplemented with tryptophan (a known precursor of violacein) between 0 and 25 °C. Violacein production was reported at a growth pH of 7.4 at 20 °C. Other growth mediums reported to produce violacein are as shown in Table 1.3.

Another excellent example of a well-studied bacterial pigment is prodigiosin. Prodigiosin(5-[(3-methoxy-5-pyrrol-2-ylidene-pyrrol-2-ylidene)methyl]-2-methyl-3-pentyl-1H-pyrrole) is a typical alkaloid compound produced as a secondary metabolite. It has a unique structure consisting of three pyrrole rings and a pyrrolyl-pyrromethene skeleton with C-4 methoxy group (Fig. 1.3). Secondary metabolites of bacterial origin include pigments, antibiotics, immunomodulators, antidiabetic and anticancer compounds (Barnett et al. 2006). Prodigiosin is produced by *Serratia marcescens* following a bifurcated biosynthesis pathway, in which mono- and bi-pyrrole precursors are obtained separately and then coupled to form the linear tripyrrole red pigment (Boger and Patel 1988) during the stationary phase of bacterial growth (Rokem and Weitzman 1987). There are several pharmaceutically relevant prodigiosins such as undecylprodigiosin, metacycloprodigiosin, roscophylin and nonylprodigiosin which are known as potential antibacterial, antimalarial, anticancer, cytotoxic and immunosuppressive agents (Song et al. 2006).

A wide variety of bacterial taxa, including Gram-negative rods such as *S. rubidaea, Vibrio gazogenes, Alteromonas rubra, Rugamonas rubra,* and

Table 1.3 Various growth media used for the production of violacein by *C. violaceum*

Growth medium	Temperature (°C)	References
Liquid medium consist of D-glucose, peptone, yeast extract	30	Rettori and Duran (1998)
Lactose broth	–	Walter (1934)
Wakimoto medium	–	Yuang et al. (2008)
Nutrient broth, Nutrient agar	25	Haisheng et al. (2008)
Soil extract agar (SEA), Soil extract broth (SEB)	15, 20, 25	Innis and Mayfield(1979)
TY medium (containes: tryptone, yeast extract and NaCl)	30	Rene'e and Kendall (2000)
RPMI1640 medium containing glutamine and antibiotics.	–	Carmen et al. (2004)
Terrific broth	37	Marlon et al. (2006)
Nutrient agar	30	DeMoss and Happle (1958)
Luria broth	28	

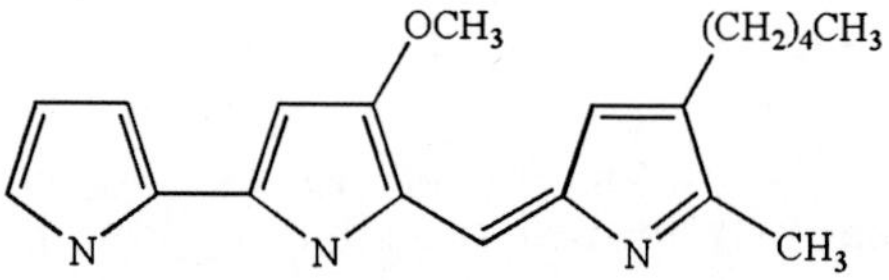

Fig. 1.3 Chemical structure of prodigiosin

Gram-positive actinomycetes, such as *Streptoverticillium rubrireticuli* and *Streptomyces longisporus* Ruber form prodigiosin and/or derivatives of this molecule. On some media *Rugamonas rubra* produces so much prodigiosin that, as the pH drops, it precipitates out within the cells and colonies change from pillar-box red to deep maroon, often with a green metallic sheen under reflected light. At this stage, most organisms in the colony are no longer viable (Maurice 2002).

Other than that, the most familiar examples of colored colonies seen in the routine soil, water and medical laboratory are those of pseudomonads such as the blue-green colonies of *Pseudomonas aeruginosa* or the yellow fluorescent colonies of *Ps. fluorescens* and related species. One example of a water soluble, nonfluorescent blue–green pigment produced by *Ps. aeruginosa* is pyocyanin which crystallises as beautiful blue needles and may have a role in respiration (Fig. 1.4). The yellow water-soluble fluorescent pigments produced by a number of *Pseudomonas* species, especially under conditions of iron limitation, are variously known as pyoverdin, pyofluorescein or simply fluorescein (Maurice 2002). Actinorhodin, a polyketide antibiotic, is produced by *Streptomyces coelicolor* A3(2), the best genetically known strain of *Streptomyces*. Its dual pigmentation i.e. red (acidic pH) and blue (alkaline pH) facilitates visual observation of its product. Actinorhodin is often used as a model for studying factors regulating the production of antibiotics. Biosynthesis of actinorhodin occurs mainly during the stationary phase in batch cultures, but may also be associated with growth depending on the medium used (Ozergin-Ulgen and Mavituna 1994). The pH of the culture medium is important as excretion of actinorhodin appears to occur

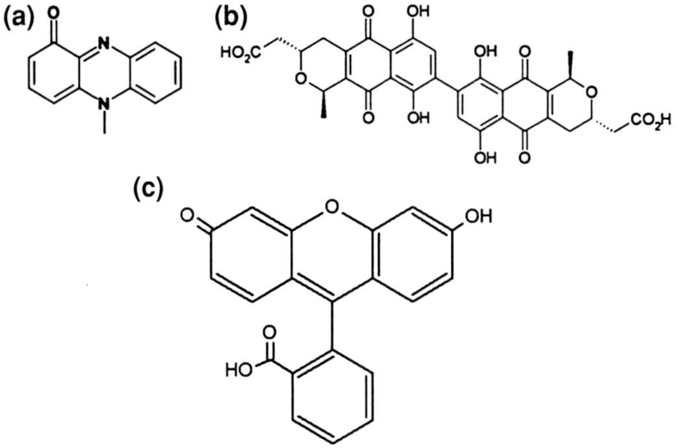

Fig. 1.4 Chemical structures for **a** pyocyanin **b** actinorhodin **c** fluorescein

exclusively at pH values above 6.70 (Wright and Hopwood 1976). Moreover, this excretion increases in complex media (Ozergin-Ulgen and Mavituna 1994) but, according to Bystrykh et al. (1996), the excreted pigment is not actinorhodin but its lactone derivative, γ-actinorhodin.

1.5.2 Fungi

Like plants, filamentous fungi synthesizes natural products (such as pigment) due to its ecological function and value (Firn and Jones 2003). Depending on the type of the compound, these natural products have different functions varying from protective mechanism against the lethal photooxidations (carotenoids) to acting as cofactors in enzyme catalysis (flavins). Besides providing the functional diversity to the host, these pigments exhibit a unique structural and chemical diversity with an extraordinary range of colors. Several characteristic non-carotenoid pigments are produced by the filamentous fungi, including quinines such as anthraquinones and naphthaquinones (Baker and Tatum 1998; Medenstev and Akimenko 1998), dihydroxy naphthalene melanin, a complex aggregate of polyketides (Butler and Day 1998) and flavin compounds such as riboflavin. *Monascus* has been reported to produce at least six major related pigments (Wang and Hesseltine 1979) which can be classified into orange (rubropunctain and monascoubrin), yellow (monascin and ankaflavin) and red (rubropunctaminea and monascorubramine). *Monascus* spp. belongs to the *Ascomycetes* group and *Monascaceae* family. The genus *Monascus* has four species namely *M. pilosus, M. purpureus, M. ruber* and *M. froridanus*, which account for the majority of strains isolated from traditional oriental food (Sabater-Vilar et al. 1999). The strain *Penicillium oxalicum* var. *armeniaca* CCM 8242, obtained from soil, produces a chromophore of the anthraquinone type. Many toxicological data are available on this red pigment; acute oral toxicity in mice (90-day subchronical toxicological study), acute dermal irritation/corrosion, acute eye irritation/corrosion, antitumor effectiveness, micronucleus test in mice, AMES test (*Salmonella typhimurium* reverse mutation assay), estimation of antibiotic activity and results of estimation of five mycotoxins (Dufossé 2006).

1.5.3 Yeast

The carotenoid imparts distinctive orange–red coloration to the animals and contributes to consumers' appeal in the market place. Astaxanthin (3,3′-dihydroxy-β,β-carotene-4,4′-dione) is widely distributed carotenoids in nature and is the principal pigment in crustaceans and salmonids. The red yeast, *Rhodotorula* synthesize carotenoid pigments consisting of torulene and torularhodin, with minute quantity of β-carotene. Animal feed supplemented with *Rhodotorula* cell mass was proven to be safe and nontoxic in animals. Its use in the nutrition of laying hens has also been documented. As β-carotene content in wild strains of *R. glutinis* is quite low, efforts have been made to increase it through strain improvement, mutation, medium optimization and manipulation of culture conditions (Sakaki et al. 2000; Tinoi et al. 2005). These studies resulted mainly in an increased yield of torulene and torularhodin, which are of minor interest.

1.5.4 Algae

Microalgae are among the fastest growing autotrophs that produce vast array of natural products including protein, enzymes, bioactive compounds and carotenoids. The unicellular microalgae, *Dunaleilla salina* is reported to produce β-carotene. The pigment was proven to have antioxidant property by quenching excessive free radicals and restoring the physiological balance. In comparison to others, *Dunaliella* has many advantages such as easier cell disruption technique than other algae because of the absence of cell wall, relatively high growth rate and higher resistant level to various environmental conditions (Pisal and Lele 2005).

1.5.5 Functions of Pigments

One of the major functions of bacterial pigment is for bacterial identification, to the extent of the incorporation of bacterial colors into its name. For example, the golden-colored pathogen *Staphylococcus aureus* (*aureus* in Latin, "golden") was named in such a way to distinguish it from the non-pigmented staphylococci i.e. *Staphylococcus alba* (*alba* in Latin, "white"). Likewise, the blue–green *Pseudomonas* species found in the lungs of patients with cystic fibrosis was given the name *aeruginosa*, which is derived from a Latin word denoting the color of copper rust. *Chromobacterium violaceum* not surprisingly, elaborates a blue–violet pigment. These hallmark phenotypes not only provide an easy nomenclature for the microorganisms, but continue to be important diagnostic clues in clinical laboratories today for the identification of microbes (Liu and Nizet 2009). Pigments have also played a role in the discovery of infectious pathogens. Some natural functions proposed for microbial pigments include protection mechanism for the microbes (against ultraviolet radiation, oxidants, extremes of heat and cold, natural antimicrobial compounds produced by other microbes), antimicrobial activities, acquisition of nutrients (such as Fe) and acquisition of energy via photosynthesis, e.g. cyanobacteria (Liu and Nizet 2009).

1.5.6 Advantages and Applications of Bacterial Pigments

One of the limitations of the use of natural dyes or pigments is the low extraction yield factors (a few grams of pigment per kg of dried raw material). Therefore, the exploitation of other biological sources such as fungi, bacteria and cell cultures offers interesting alternatives since appropriate selection, mutation or genetic engineering techniques are likely to improve significantly the pigment production yields with respect to wild organisms (Mapari et al. 2005). Microbial pigments from bacterial origins offer the advantage in terms of production, compared to pigments extracted from vegetables or animals, due to its simple cell and fast culturing technique. This will allow continuous bioreactor operation which is an important factor during industrial operation (Hendry and Houghton 1997; Babitha 2009). Besides, pigments of higher organisms such as animal, plant and fungal may be less suitable for industrial exploitation due to the structural complexity of the pigment-bearing tissue and pigment formation only at critical points of development within a complex life cycle (Hendry and Houghton 1997). For example, pigment that function as attractants in sexual reproduction may be formed only after completion of other aspects of life cycle. They may not then be amenable to exploitation through manipulation (Hendry and Houghton 1997). Another advantage of using bacterial pigments is its abundance in nature and easy propagation where only minimal medium is required because bacteria have an amazing ability to utilize cheap C and N sources to produce valuable low- and high-molecular-weight metabolites (Demain 1980). Natural raw materials and by-products of industrial processes have been proven as useful substituents for the expensive culture media during fermentation processes, which are extremely significant as culture media normally constitutes between 38–73% of the total production cost. The bacterial pigment can also be extracted using simple liquid–liquid extraction technique where the solvent used can be recovered for subsequent use, hence minimizing the operating cost.

1.5.6.1 Application as Food Colorant

Streptomyces coelicolor synthesize blue pigments which are stable to light and heat, resistant to oxidants and reducers under acidic conditions and to reducers under alkaline conditions. These characteristics makes it a good candidate for the food processing industry as an additive such as to produce colorful beverages and cakes (Zhang et al. 2006). *Bradyrhizobium* sp. strain was described as a cantha-xanthin (4,4′-diketo-β-carotene) producer (Lorquin et al. 1997). This carotenoid pigment, canthaxanthin, has been used in aquaculture feed for many years in order to impart the desired flesh color in farmed salmonids. While *Brevibacterium aurantiacum* sp. nov. is of particular interest as this microorganism is found in the rind of red- smear ripened soft cheeses, hence its long association with human consumption (Galaup et al. 2005; Dufossé et al. 2005). Other than that, the yellow pigment known as zeaxanthin or 3,3′-dihydroxy-β-carotene, produced by *Flavobacterium* sp. can be used as an additive in poultry feeds to strengthen the yellow color of the skin of animals of this kind or to accentuate the color of the yolk of their eggs (Alcantara and Sanchez 1999). Other bacterial pigments already in use as natural food colorants or with high potential in this field includes astaxh-antin (pink–red pigment) from *Agrobacterium aurantiacum* (Yokoyama et al. 1994) and *Paracoccus carotinifaciens* (Tsubokura et al. 1999), rubrolone (red pigment) from *Streptomyces echinoruber*, zeaxanthin (yellow pigment) from *Flavobacterium* sp. (Shepherd et al. 1976) and *Paracoccus zeaxanthinifaciens* (Dufossé 2006).

1.5.6.2 Application in Pharmaceutical Industry

The genus, *Streptomyces* or *Serratia* can produce a red substance of pyrrolylpy-romethene skeleton, which is one of following substances: prodigiosin, metacyc-leprodigiosin, prodigiosin, desmethoxy prodigiosin, and prodigiosin 25-C. These substances have been known to have an antibiotic and antimalarial effect, espe-cially prodigiosin 25-C that shows immunosuppressing activity (Kim et al. 2003). Immunosuppressive activity of prodigiosin was first described by Nakamura and coworkers in 1989. These researches showed the presence of prodigiosin and metacycloprodigiosin in culture broth of *Serratia* and observed selective inhibition of polyclonal proliferation of T-cells as compared to that of B-cells. Besides that, the cytotoxic potency of prodigiosin has also been investigated in the standard 60 cell line panels of human tumor cells derived from lung, colon, renal, ovarian, brain cancers, melanoma and leukemia. Inhibition of cell proliferation as well as induction of cell death has been observed in these cell lines. In vitro anticancer activity has also been reported for different prodigiosin analogs and synthetic indole derivative of prodigiosin (Pandey et al. 2007). The antiproliferative and cytotoxic effects of prodigiosin have been observed not only in cultured tumor cell lines but also in human primary cancer cells from B-cell chronic lymphocytic leukemia patients (Campas et al. 2003). The use of prodigiosin for treating diabetes mellitus has also been reported by Hwanmook et al. (2003) where prodigiosin was found to be an active component for preventing and treating diabetes mellitus. Violacein is produced by several bacterial species, including the Gram-negative species *Chromobacterium violaceum*, *Janthinobacterium lividum*, *Pseudoalteromonas luteoviolacea*, *Ps.* sp 520P1 and *Ps.* sp. 710P1 (Rettori and Duran 1998; Pantanella et al. 2007; Yada et al. 2007). Violacein was reported to have antiprotozoan (Matz et al. 2004; Leon et al. 2001), anticancer (Ferreira et al. 2004; Kodach et al. 2006), antiviral (Andrighetti-Fröhner et al. 2003), antibacterial (Sánchez et al. 2006; Lichstein and van De Sand 1946; Nakamura et al. 2003) and antioxidant activities (Konzen et al. 2006). These characteristics provide the possible applications of violacein for therapeutic purposes (Richard 1993).

1.5.6.3 Application in Textile Dyeing

Currently used colorants are almost exclusively made from non-renewable resources such as fossil oil. The production of the synthetic colorants is economically efficient and technically advanced with colors covering the whole color spectrum. However, synthetic colorants are facing the following challenges namely dependence on non-renewable oil resources and sustainability of current operation, environmental toxicity and health concerns. Thus, biosynthesis of pigments through fermentation processes can serve as major chromophores for further chemical modifications, which could lead to colorants with a broad spectrum of colors (Hobson and Wales 1998). Besides, some natural colorants, especially anthraquinone type compounds, have shown remarkable antibacterial activity in addition to providing bright colors (Frandsen et al. 2006), which could serve as functional dyes in producing colored antimicrobial textiles. Alihosseini et al. (2008) characterized the bright red pigment prodigiosin from *Vibrio* spp. and suggested that it could be used to dye many fibers including wool, nylon, acrylics and silk. Yusof (2008) reported the capability of using pigment from *Serratia marcescens* to color five types of fabric namely acrylic, polyester microfiber, polyester, silk and cotton using tamarind as mordant. However, the dyeing performances are different, depending on the types of fiber. From the colorfastness testing, the dyed fabrics also have the ability to maintain its color under several external conditions such as perspiration, washing, and rubbing/crocking. Similar textile-dyeing ability was also reported for *Janthinobacterium lividum* (Shirata et al. 2000), an isolate from wet silk thread. The purple pigment from this bacterium gave good color tone when applied on silk, cotton and wool (bluish-purple, all natural fibers), and nylon and vinylon (dark blue, both synthetic fibers). Dyeing was performed by a simple procedure consisting of either dipping in the pigment extract or boiling with the bacterial cells. Color variation was achieved by changing the dipping time and the temperature of the dye bath. The color fastness of the dyed material was similar to vegetable pigment-dyed materials, however with lower lightfastness. The pigment displayed an antimicrobial activity against phytopathogenic fungi like *Rosellinia necatrix* which causes white root rot of mulberry (Shirata et al. 2000).

1.6 Growth Medium

Microbial growth medium can be classified into three broad categories namely consistency, nutritional component and functional use. *Consistent media* can be either in liquid, solid, semi-solid or biphasic forms: (1) *Liquid medium*, normally referred to as broth, is normally used in bottles or flasks. Liquid media allows easy and fast cultivation of bacteria in great cell number as well as minimizing the impact of any bacterial growth inhibitors present via dilution effect. However, properties of bacteria are not readily visible with possible presence of bacterial co-cultures. Examples for liquid medium include nutrient broth, skimmed milk and peptone solution, (2) *Solid medium* is easily obtained via the addition of solidifying agar into liquid medium. Agar is composed of two long-chain polysaccharides (70% agarose and 30% agarapectin) with a melting point of 95 °C. Without significant nutritional content (even though it can be source for Ca and simple organic ions), agar is not hydrolyzed by most bacteria and is usually free from growth-promoting or growth-retarding substances. Some examples for solid medium are as follows; trypticase soy agar, potato slices and coagulated blood serum, (3) *Semi-solid* agar is prepared by using a smaller amount of agar compared to solid medium and is fairly useful in demonstrating bacterial motility. One example is the cystine trypticase agar medium. Classification based on *nutritional component* includes two main groups i.e. non-synthetic and synthetic media. A non-synthetic media (complex media or undefined media) is a medium in which the exact chemical composition of each of the constituents is not known with certainty. Amongst the most widely used non-synthetic media include Nutrient broth, Potato dextrose agar, Luria-Bertani broth, Oatmeal agar and Soil extract agar. Non-synthetic media usually consist of digests of casein, soybean, beef, yeast cells or any other highly nutritious material without defined substances. A synthetic media (defined media or minimal media) is medium in which only pure chemicals in definite concentrations are used as it provides only the exact nutrients (including any growth factors) needed by the organism for growth. The use of defined media requires the investigator to know the exact nutritional requirements of the organisms in question and is particularly useful in metabolic studies. Bacteria that are able to grow in minimal media are termed as non-fastidious while those that require extra nutrients are said to be fastidious. Classification of media based on functional use or application can be divided into cultivation, storage, enrichment, differential, assay and transport media. *Cultivation media* is used for general cultivation of bacteria (nutrient broth, Luria-Bertani etc.), *storage media* where bacteria are stored in "stock culture" condition for longer periods to provide a source of viable cultures for future use; *enrichment media* in which the nutrient content is adjusted as to enhance the growth of certain bacterium in a mixed bacterial population. This approach is extremely useful during the isolation of pure bacterial colonies from environmental samples or mixed bacterial population in the laboratory. Some examples on the application of enrichment media are as given in Table 1.4.

Table 1.4 Applications of enrichment media

Incubation Condition	Main ingredients (electron donor)	Electron acceptor	Organisms enriched	Isolated from
Aerobic	NH_4^+	O_2	*Nitrosomonas*	Soil, mud, sewage effluent
	Lactate + NH_4^+	O_2	*Pseudomonas fluorescens*	Soil, mud, lake sediments, decaying vegetation
	Ethanol (4%) + Yeast extract (1%), pH 6.0	O_2	*Acetobacter, Gluconobacter*	
Anaerobic	Organic acids	KNO_3 (1%)	*Pseudomonas* (denitrifying species)	Soil, mud, lake sediments
	Acetate, propionate, butyrate	Na_2SO_4	Fatty-acid-oxidizing sulfate reducers	Sewage digester sludge, soil, mud, lake sediments
	S, $S_2O_3^{2-}$	NO_3^-	*Thiobacillus denitrificans*	Soil, mud, lake sediments

Differential media is used to determine differential reactions which are useful to give a presumptive idea on the identity of bacterial species. One example of a differential media is the blood agar medium which is used to identify the presence of haemolytic bacteria. A hemolytic bacterium would create certain clearing zones on the agar plate from the hemolysis of red blood cells while bacteria with non-hemolytic features would not give such clearing zones. *Assay media* are those with prescribed compositions that have a direct influence on the formation of certain enzymes, toxins, antibiotics and other products. This would allow qualitative and quantitative determination of one or more components, both primary and secondary metabolites. Transport media prevents the overgrowth of contaminating organisms or commensals in fresh samples from the point of sampling to the laboratory. Transport media such as Stuart's and Amie's, Cary Blair, Venkatraman Ramakrishnan and Sach's buffered glycerol saline would prevent desiccation of specimen, maintain existing pathogen to commensal ratio and inhibit overgrowth of unwanted bacteria. Some of these media (Stuart's & Amie's) are semisolid in consistency.

1.7 Collection of Samples

Sample collection techniques are normally fabricated to ensure efficient and successful organism detection and enumeration in the laboratory, hence application of proper sample collection and storage technique are of paramount importance. Factors such as improperly prepared sampling containers, inadvertent contamination of samples, inappropriate sample storage conditions and excessive transport time can directly affect detection and quantitative validity. Sampling container for microbiological examination should consist of either nonreactive borosilicate

Table 1.5 Features of sample collection from different environmental sources

Source	Features
Potable water	• source: drinking water distribution line, well • sampling from tap connected directly to the main line • let water run to waste for 2–3 min (clearing of service line) prior to collecting sample (slower flow, without splashing) • sampling apparatus: ZoBell J–Z sampler
Raw water supply	• source: river, stream, lake, reservoir, spring, shallow well • undesirable to take samples too near the bank or too far from the point of draw off, or at the depth above or below the point of draw off
Surface waters	• source: recreational water, water supply, waste treatment control, effluent discharging (i.e. critical sites) • frequency—ranging from hourly (effluent discharging) to seasonal (recreational waters)
Sediments and biosolids	• source: water supply reservoirs, lakes, rivers, coastal waters • sediments may provide a stable index of the general quality of the overlying water • biosolids with less than 7% total solids should be collected (>7% solids require a finite shear stress) • sampling apparatus: stainless steel fitted with sterile plastic bags
Non-potable samples (manual sampling)	• source: river, stream, lake, reservoir • sample collected in flowing stream direction (may create artificial current) • avoid contact with stream or bank bed to avoid water fouling

glass, plastic bottles or plastic bags which have been thoroughly rinsed using deionized water prior to autoclaving. Aseptic techniques should be applied throughout the sampling process. Some of the features for collecting samples from different environmental sources as described in Table 1.5.

For *sample preservation and storage*, determination on the microbiological content of a sample should be carried out as soon as (in situ) the sampling has been carried out. For drinking water, the sampling containers should be kept at less than 10 °C during transportation to the laboratory (i.e. preserve in iced container). Determination for coliform bacteria must be carried out less than 30 h from the time of sampling while 8 h for heterotrophic plate counts. Similar condition (less than 10 °C) also applied to non-potable water with a maximum transport time of 6 h. For other water types, samples must be kept at less than 10 °C with maximum holding time of 24 h.

1.8 Characterization of Microorganisms

A variety of techniques exist for the analysis of microorganisms. The techniques can generally be divided into morphological, biochemical, physiological and molecular characterization. *Morphological* characterization includes plate observation on the features of bacterial colonies formed, microscopic (light and electron), Gram characteristics as well as various staining procedures i.e. to detect the presence of specific bacterial embodiment such as endospore, capsule or flagella. *Biochemical* characterization allows preliminary identification of microbes via examining for the presence of certain enzymes namely protease, DNase, lipase, lecithinase, oxidase and catalase. Initially, various specific ready- made medium such as Himedia laboratories were used for the enzymatic assay. However, the availability of rapid test kit such as API 20E, API 20NE (manual) and the automated (BIOLOG^{TM}, Vitek.R^{TM}, Phoenix^{TM} and Microscan.R^{TM}) systems enable quick, efficient and reliable method for bacterial identification. The API 20E system for example, consists of a plastic strip of 20 individual, miniaturized tests tubes (cupules), each containing a different reagent used to determine the metabolic capabilities, and, ultimately, the genus and species of enteric bacteria in the family *Enterobacteriaceae*. The Biolog Identification System is a bacterial identification method that was established based on the exchange of electrons generated during respiration. This system tests the ability of a microorganism to oxidize a panel of 95 different carbon sources. Biolog's microplates use redox chemistry to colorimetrically indicate respiration of live cell suspensions. All wells are initially colorless. When a chemical in a well is oxidized, there is a burst of respiration and the cells reduce a tetrazolium dye, forming a purple color. A reference well contains no carbon source. The test yields a pattern of purple wells which constitutes a "metabolic fingerprint" of the organism. *Physiological* characterization includes determination of optimum pH, temperature for growth, efficiency in substrate utilization, growth profile under enriched condition, and metal tolerance as well as sensitivity for different antibiotics. However, these techniques are not absolutely definitive for the identification of microbes since many organisms have similar phenotypic characteristics, which make accurate identification very difficult. In addition, large concentrations of microbes are necessary for analysis. This prompted the shift on bacterial characterization/identification towards *molecular* characterization which was based on 16S rDNA sequence analysis. The steps for the characterization included the genomic DNA isolation followed by PCR amplification of universal 16S rDNA gene. However, sample purification and DNA isolation prior to PCR analysis prove to be both time-consuming and cumbersome. Other techniques used for the identification of bacteria include flow cytometry (bacterial number/concentration), mass spectrometry (cell components molecular fingerprint, a series of molecular mass/charge ratio intensities), Fourier transform infrared (FT–IR), Raman spectroscopy (fingerprint of functional groups present on the microbes, less sensitive than mass spectrometry), fluorescence spectrometry, microelectrophoresis (movement of microbes—naturally charged, in a direct-current electric field), laser desorption ionization time-of-flight mass spectrometry, MALDI-TOF–MS (matrix assisted analysis of either cellular protein extracts or whole cells) and two-dimensional ultracentrifuging (extract, separate, and purify microbes including viruses from body fluids or homogenized tissue).

References

Alcantara S, Sanchez S (1999) Influence of carbon and nitrogen sources on *Flavobacterium* growth and zeaxanthin biosynthesis. J Ind Microbiol Biot 23:697–700

Alihosseini F, Ju KS, Lango J, Hammock BD, Sun G (2008) Antibacterial colourants: Characterization of prodiginines and their applications on textile materials. Biotechnol Progr. 24:742–747

Alves SP, Brum DM, de Andrade ECB, Netto ADP (2008) Determination of synthetic dyes in selected foodstuffs by high performance liquid chromatography with UV-DAD detection. Food Chem 107:489–496

Andrighetti-Fröhner CR, Antonio RV, Creczynski-Pasa TV, Barardi CRM, Simões CMO (2003) Cytotoxicity and potential antiviral evaluation of violacein produced by *Chromobacterium violaceum*. Memórias do Instituto Oswaldo Cruz. 98:843–848

Babitha S (2009) Chapter 8: Microbial pigments.In: Nigam PS , Pandey A (eds), Biotechnology for agro-industrial residues. Springer, Dodrdrecht, pp. 147–162

Babitha S, Sandhya C, Pandey A (2004) Natural Food Colourants. Appl Bot Abstracts 23(4):258–266

Baker RA, Tatum JH (1998) Novel anthraquinones from stationary cultures of *Fusarium oxysporum*. J Fermen Bioeng 85:359–361

Barnett JR, Sarah Miller and Pearce E (2006) Colour and art: A brief history of pigments. Optics Las Tech 38:445-453

Boger D, Patel M (1988) Total synthesis of prodigiosin, prodigiosene, and, desmethoxyprodigiosin: Diels-Alder reactions of heterocyclic azadienes and development of an effective palladium (II)-promoted 2, 29-bipyrrole coupling procedure. J Org Chem 53:1405–1415

Butler MJ, Day AW (1998) Fungal Melanins: A Review. Canadian J Microbiol 44:1115–1136

Buxbaum G (1993) Industrial Inorganic Pigments. Weinheim: VCH. p 85

Bystrykh LV, Ferna′ndez-Moreno MA, Herrema JK, Malpartida F, Hopwood DA, Dijkhuizen L (1996) Production of actinorhodin-related 'blue pigments' by *Streptomyces coelicolor* A3(2). J Bacteriol 178:2238–2244

Campas C, Dalmau M, Montaner B, Barragan M, Bellosillo B, Colomer D (2003) Prodigiosin Induces Apoptosis of B and T Cells from B-Cell Chronic Lymphocytic Leukemia. Leukemia 17:746–750

Cao W, Chen W, Sun S, Guo P, Song J, Tian C (2007) Investigating the antioxidant mechanism of violacein by density functional theory method. J Theochem 817:1–4

Carmen VF, Carina LB, Henri HV, Giselle ZJ, Nelson D, Maikel PP (2004) Molecular mechanism of violacein-mediated human leukemia cell death. Blood 104:1459–1464

Christie RM (2001) Colour chemistry.Royal Society of Chemistry, United Kingdom, pp 12–44,

Chung K-T, Stevens SV, Cerniglia CE (1992) The reduction of azo dyes by the intestinal microflora. Crit Rev Microbiol 18:175–190

Clinton SK(1998) Lycopene chemistry, biology, and implications for human health and disease. Nutrition reviews

Cristea D, Vilarem G (2006) Improving light fastness of natural dyes on cotton yarn. Dyes Pigments 70:238–245

Daniel MM (1986) Handbook of U.S. colorant for food, drugs, and cosmetics. 2nd edn. Wiley-Interscience , USA

Demain AL (1980) Microbial production of primary metabolites. Naturwissenschaften 67: 582–587

Demming-Adams B (2002) Antioxidants in photosynthesis and human nutrition, Science: w.w.adam III .298

DeMoss RD, Happle NR (1958) Nutritional requirements of *Chromobacterium violaceum*. Department of Microbiology, University of Illinois, Urbana 77:137–141

Dufossé L (2006) Microbial production of food grade pigments. Food Technol Biotech 44(3):313–321

Dufossé L, Galaup P, Carlet E, Flamin C, Valla A (2005) Spectrocolorimetry in the CIE L*a*b* color space as useful tool for monitoring the ripening process and the quality of PDO red-smear soft cheeses. Food Res Int 38:919–924

Durán N, Menck CFM (2001) *Chromobacterium violaceum*: A review of pharmacological and industrial perspectives. Crit Rev Microbiol 27:201–222

Eichlerová I, Homolka L, Nerud F (2006) Synthetic dye decolourization capacity of white rot fungus *Dichomitus Squalens*. Bioresource Technol 97:2153–2159

Ferreira CV, Bos CL, Versteeg HH, Justo GZ, Durán N, Peppelenbosch MP (2004) Molecular mechanism of violacein-mediated human leukemia cell death. Blood 104:1459–1467

Firn RD, Jones CG (2003) Natural products - a simple model to explain chemical diversity. Nat Prod Rep 20:382–391

Forgacs E, Cserháti T, Oros G (2004) Removal of synthetic dyes from wastewaters: A review. Environ Int 30:953–971

Frandsen RJN, Nielsen NJ, Maolanon N, Sorensen JC, Olsson S, Nielsen J, Giese H (2006) The biosynthetic pathway for aurofusarin in *Fusarium graminearum* reveals a close link between the naphthoquinones and naphthopyrones. Mol Microbiol 61:1069–1080

Galaup P, Flamin C, Carlet E, Dufossé L (2005) HPLC analysis of the pigments produced by the microflora isolated from the'Protected Designation of Origin' french red-smear soft cheeses munster, epoisses, reblochon and livarot. Food Res Int 38:855–860

Goodwin TW (2002) Chemistry and biochemistry of plant pigments. Academic Press, London

Haisheng W, Peixia J, Yuan L, Zhiyong R, Ruibo J, Xin-Hui X, Kai L, Dong W (2008) Optimization of culture condition for violacein production by a new strain of Duganella sp B2. Biochem Eng J 44:119–124

Hendry GAF, Houghton JD (1996) Natural food colorants, 2nd edn. Blackies Academic & Professional: imprint of Champan & Hall, Great Britain

Hendry GAF, Houghton JD (1997) Natural food colourants. Blackie & Son, Glasgow

Hiromichi K, Fumitaka H (2002) An approach to estimate the chemical structure of melanoidins. International Congress Series 1245:3–7

Hobson DK, Wales DS (1998) Green colourants. JSDC 114:42–44

Hu TL (2001) Kinetics of azoreductase and assessment of toxicity of metabolic products from azo dye by *Pseudomonas Luteola*. Water Sci Technol 43:261–269

Hwanmook K, Sangbae H, Changwoo L , Kihoon L, Sehyung P, Youngkook K (2003) Use of prodigiosin for treating diabetes mellitus, US Patent 6638968

Innis WE, Mayfield CI (1979) Effect of temperature on violacein in a psychotropic *Chromobacterium* from lake Ontario sediment. Canada Microb Ecol 5:51–56

Jaworska M, Szulińska Z, Anuszewska MWE (2005) Separation of synthetic food colourants in the mixed micellar system application to pharmaceutical analysis. J Chromatogr A 1081:42–47

Kamel MM, El-Shishtawy RM, Yussef BM, Mashaly H (2005) Ultrasonic assisted dyeing III. Dyeing of wool with lac as a natural dye. Dyes Pigments 65:103–110

Kapadia GJ, Tokuda H, Sridhar R, Balasubramanian V, Takayasu J, Bu P, Enjo F, Takasaki M, Konoshima T, Nishin H (1998) Cancer chemopreventive activity of synthetic colourants used in foods, pharmaceuticals and cosmetic preparations. Cancer Lett 129:87–95

Kim H, Han SB, Lee OW, Lee K, Park S, Kim Y (2003) Use of prodigiosin for treating diabetes mellitus, US patent 6,638,968 B1.

Kodach LL, Bos CL, Durán N, Peppelenbosch MP, Ferreira CV, Hardwick JCH (2006) Violacein synergistically increases 5-fluorouracil cytotoxicity, induces apoptosis and inhibits aktmediated signal transduction in human colourectal cancer cells . Carcinogenesis 27:508–516

Konzen M, de Marco D, Cordova CAS, Vieira TO, Antônio RV, Creczynski-Pasa TB (2006) Antioxidant properties of violacein: Possible relation on its biological function . Bioorgan Med Chem 14:8307–8313

Leon LL, Miranda CC, De Souza AO, Durán N (2001) Antileishmanial activity of the violacein extracted from *Chromobacterium violaceum*. J Antimicrob Chemoth 48:449

Lichstein HC, van De Sand VF (1946) The antibiotic activity of violacein, prodigiosin, and phthiocol . J Bacteriol 52:145–146

Liu GY, Nizet V (2009) Colour me bad: Microbial pigments as virulence factors. Trends Microbiol 17(9):406–413

Lorquin J, Molouba F, Dreyfus BL (1997) Identification of the carotenoid pigment canthaxanthin from photosynthetic *Bradyrhizobium* strains . Appl Environ Microb 63:1151–1154

Mapari SAS, Nielsen KL, Larsen TO, Frisvad JC, Meyer AS, Thrane U (2005) Exploring fungal biodiversity for the production of water-soluble pigments as potential natural food colourants. Curr Opin Biotech 16:231–238

Marlon K, Daniela DM, Clarissa ASC, Tiago OV, Regina VA, Tina BCP (2006) Antioxidant properties of violacein: possible relation on its biological function. Bioorgan Med Chem. 14:8307–8313

Matz C, Deines P, Boenigk J, Arndt H, Eberl L, Kjelleberg S, Jürgens K (2004) Impact of violacein-producing bacteria on survival and feeding of bacterivorous nanoflagellates. Appl Environ Microb 70:1593–1599

Maurice, M (2002) Bacterial Pigments. Microbiologist. 10-12.

Medenstev AG, Akimenko VK (1998) Naphthoquinone metabolites of the fungi. Phytochemistry 47:935–959

Melo PS, Justo GZ, De Azevedo MBM, Durán N, Haun M (2003) Violacein and its beta-cyclodextrin complexes induce apoptosis and differentiation in HL60 cells. Toxicology 186:217–225

Nagia FA, EL-Mohamedy RSR (2007) Dyeing of wool with natural anthraquinone dyes from *Fusarium oxysporum*. Dyes Pigments 75:550–555

Nakamura A, Magae J, Tsuji RF, Yamasaki M, Nagai K (1989) Dyeing of wool with natural anthraquinone dyes from *Fusarium oxysporum*. Transplantation 47:1013–1016

Nakamura Y, Asada C, Sawada T (2003) Production of antibacterial violet pigment by psychrotropic bacterium RT102 strain. Biotechnol Bioproc E 8:37–40

Ozergin-Ulgen K, Mavituna F (1994) Comparison of the activity of immobilized and freely suspended *Streptomyces coelicolor* A3(2). Appl Microbiol Biot 41:197–202

Pandey R, Chander R, Sainis KB (2007) Prodigiosins: A novel family of immunosuppressants with anticancer activity. Indian J Biochem Bio 44:295–302

Pantanella F, Berlutti F, Passariello C, Sarli S, Morea C, Schippa S (2007) Violacein and biofilm production in *Janthinobacterium lividum*. J Appl Microbiol 102:992–999

Pelz A, Wieland K-P, Putzbach K, Hentschel P, Albert K, Gotz F (2005)Structure and biosynthesis of staphyloxanthin from *Staphylococcus aureus*. J Biol Chem 280:32493–32498

Pisal DS, Lele SS (2005) Carotenoid production from microalgae, *Dunaliella salina*. Indian J Biotechnol 4:476–483

Rattanaphani S, Chairat M, Bremner BJ and Rattanaphani V (2007). An adsorption and thermodynamic study of lac dyeing on cotton pretreated with chitosan. Dyes Pigments. 72:88–96

Rene'e SB, Kendall MG (2000) Extraction of violacein from *Chromobacterium violaceum* provides as new quantitative bioassay for N-acyl homoserine lactone autoinducers. J Microbiol Meth. 40:47–55

Rettori D, Duran N (1998) Production, extraction and purification of violacein: An antibiotic pigment produced with *Chromobacteria violaceum*. World J Microb Biot. 14:685–688

Richard C (1993)*Chromobacterium violaceum*, opportunist pathogenic bacteria in tropical and subtropical regions. Bull Soc Path Exot 86:169–173

Richella M, Bortlik K, Liardet S, Hager C, Lambelet P, Baur M et al (2002) A food-based formulation provides lycopene with the same bioavailability to humans as that from tomato paste. J Nutr 132:404–408

Rokem J, Weitzman P (1987) Prodigiosin formation by *Serratia marcescens* in chemostat. Enzyme Microb Tech 9:153–5.

Rowe KS, Rowe KJ (1994) Synthetic food colouring and behavior: A dose response effect in a double-blind, placebo-controlled, repeated-measures study. J Pediatr 125:691–698

Sabater-Vilar M, Maas RFM, Fink-Gremmels J (1999) Mutagenicity of commercial *Monascus* fermentation products and the role of citrinin contamination. Mutat Res 444:7–16

Sakaki H, Nakanishi T, Satonaka KY, Miki W, Fujita T, Komemushi S (2000) Properties of a high-torularhodin mutant of *Rhodotorula Glutinis* cultivated under oxidative stress. J Biosci Bioeng 89:203–205

Sánchez C, Braña AF, Méndez C, Salas JA (2006) Reevaluation of the violacein biosynthetic pathway and its relationship to indolocarbazole biosynthesis. Chem Bio Chem. 7:1231–1240

Shepherd D, Dasek J, Suzanne M , Carels C (1976) Production of zeaxanthin. US Patent 3,951,743

Shirata A, Tsukamoto T, Yasui H, Hata T, Hayasaka S, Kojima A, Kato H (2000)Isolation of bacteria producing bluish-purple pigment and use for dyeing. Japan Agric. Res. Quart 34:131–140

Song MJ, Bae J, Lee DS, Kim CH, Kim JS, Kim SW, Hong SI (2006) Purification and characterization of prodigiosin produced by integrated bioreactor from *Serratia* sp. KH-95. J Biosci Bioeng 101(2):157–161

Timberlake CF, Henry BS (1986) Plant pigments as natural food colours. Endeavour 10(1):31–36

Tinoi J, Rakariyatham N, Deming RL (2005)Simplex optimization of carotenoid production by *Rhodotorula glutinis* using hydrolyzed mung bean waste flour as substrate. Process Biochem 40:2551–2557

Tsubokura A, Yoneda H, Mizuta H (1999) *Paracoccus carotinifaciens* sp nov., a new aerobic Gram-negative astaxanthin-producing bacterium. Int J Syst Bacteriol 49:277–282

Turner GPA (1993) Introduction to paint chemistry and principles of paint technology, 3rd edn. Chapman and Hall, London

Ullmann F (1985) Ullmann encyclopedia of industrial chemistry ,2nd edn, Edit,VCH Weinheim

Walter CT (1934) The pigment of *Bacillus violaceus*. J Bacteriol 29(3):223–227

Wang HL, Hesseltine CW (1979) Mold-modified foods. In: Peppler HJ, Perlman J (eds) Microbial Technology. Academic Press, New York, pp 95–129

Wieland B, Feil C, Gloria-Maercker E, Thumm G, Lechner M, Bravo J-M, Poralla K, Gotzl F (1994) Genetic and biochemical analyses of the biosynthesis of the yellow carotenoid 4, 40-diaponeurosporene of *Staphylococcus aureus*. J Bacteriol 176:7719–7726

Wong PK, Yuen PY (1996) Decolourization and biodegradation of methyl red by *Klebsiella pneumoniae* RS-13. Water Res 30:1736–1744

Wright LF, Hopwood DA (1976) Actinorhodin is a chromosomally-determined antibiotic in *Streptomyces coelicolor* A3(2). J General Microb 96:289–297

Yada S, Wang Y, Zou Y, Nagasaki K, Hosokawa K, Osaka I, Arakawa R, Enomoto K (2007) Isolation and characterization of two groups of novel marine bacteria producing violacein. Mar Biotechnol 10:128–132

Yokoyama A, Izumida H, Miki W (1994) Production of astaxanthin and 4-ketozeaxanthin by the marine bacterium, *Agrobacterium aurantiacum*. Biosci Biotech Bioch 58:1842–1844

Yuang L, Liyan W, Yuan X, Chong Z, Xin-Hui X, Kai L, Zhidong Z, Yong L, Guifeng Z, Jingxiu B, Zhiguo S (2008) Production of violet pigment by a newly isolated psychrotrophic bacterium from a glacier in Xinijiang, China. Biochem Eng J. 43:135–141

Yusof, NZ (2008) Isolation and applications of red pigment from *Serratia marcescens*. Universiti Teknologi Malaysia: BSc thesis.

Zhang H, Zhan J, Su K, Zhang Y (2006) A kind of potential food additive produced by *Streptomyces coelicolor*: Characteristics of blue pigment and identification of a novel compound, λ-actinorhodin. Food Chem 95:186–192

Zollinger H (1991)Colour chemistry: Syntheses, properties, and applications of organic dyes and pigments, 3rd edn. Wiley-VCH, Zurich

Chapter 2
Isolation of Pigment-Producing Bacteria and Characterization of the Extracted Pigments

Abstract Bacteria produce pigments for various reasons and it plays an important role. Some bacteria such as cyanobacteria have phycobilin pigments to carry out photosynthesis. Other example for pigment-producing bacterial strains includes *Serratia marcescens* that produces prodigiosin, *Streptomyces coelicolor* (prodigiosin and actinorhodin), *Chromobacterium violaceum* (violacein) and *Thialkalivibrio versutus* (natronochrome and chloronatronochrome). These bacteria can be isolated/cultured/purified from various environmental sources such as water bodies, soil, on plant, in insects and in man or animal. Various growth mediums can be used to isolate different types of bacteria. However, due to the high cost of using synthetic medium, there is a need to develop new low cost process for the production of pigments as well as during the isolation procedure. The use of agro-industrial residues for example, would provide a profitable means of reducing substrate cost. Pigment produced by the bacteria can be isolated using solvent extraction. These pigments can be further purified and characterized for physical and chemical characteristics using various instrumental-based analytical techniques such as TLC, UV–vis Spectroscopy, FTIR, ESI–MS, NMR HPLC and Gel Permeation Chromatography.

Keywords Pigment · Bacteria · Isolation · Characterization · NMR, HPLC · Extraction · Medium

2.1 Growth Medium

Nutrient broth (8 g L^{-1}, Merck) and Nutrient agar (20 g L^{-1}, Merck) were used as growth medium. The mediums were sterilized by autoclaving at 121 °C, 103.42 kPa for 15 min where the agar was allowed to harden followed by incubation for 24 h at 30 °C to ensure that it was free from contamination.

W. A. Ahmad et al., *Application of Bacterial Pigments as Colorant,*
SpringerBriefs in Molecular Science, DOI: 10.1007/978-3-642-24520-6_2,
© The Author(s) 2012

Table 2.1 Characteristics of water and soil samples collected from Brackishwater Aquaculture Research Centre, Johor

Sample	Sampling location/Type of sand	color	°C	pH
W1	Recycle water tank (*siakap* sp.)	Greenish Brown	29.7	6.24
W2	Recycle water tank (*tilapia* sp.)	Cloudy grey	31.9	6.24
W3	*Rotifer* breeding tank	Clear	28.3	6.22
W4	Fish breeding tank	Clear	32.1	6.29
W5	Water point source (from sea)	Cloudy grey	32.4	6.21
W6	Shrimp pond	Clear	28.3	6.14
W7	Organic waste tank (red *tilapia* sp.)	Cloudy grey	29.5	6.24
W8	Organic waste tank	Cloudy grey	28.5	6.18
S1	Clay	Orange	n/a	7.12
S2	Small sand	Grey	n/a	6.02
S3	Small sand	Grey	n/a	6.11
S4	Clay	Yellow	n/a	5.57
S5	Mixture of sand and clay	Dark green	n/a	6.23
S6	Clay	Orange	n/a	6.01
S7	Small sand	Grey	n/a	7.11
S8	Clay	Orange	n/a	6.14
S9	Mixture of sand and clay	Dark green	n/a	4.93
S10	Clay	Orange	n/a	6.22
S11	Small sand	Black	n/a	5.24
S12	Mixture of sand and clay	Dark green	n/a	6.11
S13	Small sand	Black	n/a	5.77
S14	Mixture of sand and clay	Dark grey	n/a	6.34
S15	Small sand	Grey	n/a	6.65
S16	Small sand	Black	n/a	5.88

2.2 Location and Techniques of Sampling

Liquid and soil samples were obtained aseptically from the Brackishwater Aquaculture Research Centre (BARC), Johor and one oil refinery facility in Port Dickson, Negeri Sembilan. A total of 16 soil samples, consisting of a mixture of clay and sand, were collected in the vicinity of the wastewater treatment pond of the oil refinery while at the Brackishwater Aquaculture Research Centre, solid samples were collected near the shrimp pond and the liquid samples, from the effluent fraction of various tanks namely fish rearing, rotifer breeding, fish breeding and organic waste collector. All sampling procedures were carried out according to Standard Methods for the Examination of Water and Wastewater (Eaton and Franson 2005). Pre-sterilized 250 mL Schott bottles were filled with liquid and soil samples and ample air space (for liquid samples), about 2.5 cm, were left to facilitate mixing, aeration and thermal expansion normally encountered during handling and transportation. The bottles were placed inside an iced-box polystyrene container during the 2–3 h transportation journey to the laboratory (to minimize indigenous microbial activity and preserve original speciation of chemicals present). Some characteristics of the water and soil samples are shown in Table 2.1.

Table 2.2 Phenotypic characteristics for some of the bacterial colonies isolated from soil and water samples obtained from BARC, Johor and one oil refinery in Port Dickson, Negeri Sembilan

Bacteria ID	Colony appearance	Colony Characteristics		
		Form	Elevation	Margin
W5a	Light-Yellow	Irregular	Flat	Erose
S6a	Pale-Yellow	Punctiform	Convex	Entire
S7a	Yellow	Punctiform	Convex	Entire
S8a	Light-yellow	Filamentous	Raised	Fillamentous
S8b	Yellow-orange	Punctiform	Convex	Entire
S8c	Cream-yellow	Circular	Convex	Entire
S8d	Yellow-orange	Punctiform	Convex	Entire
S10a	Yellow	Punctiform	Convex	Entire
S10c	Cream-yellow	Punctiform	Convex	Entire
S1a	Violet	Punctiform	Raised	Entire
S1b	Opaque-yellow	Circular	Umbonate	Entire
S2d	Light-yellow	Punctiform	Raised	Entire
S5e	Pale-yellow	Punctiform	Raised	Entire
S6c	Pale-yellow	Punctiform	Umbonate	Undulate
S7a	Yellow	Punctiform	Convex	Entire
S10a	Light-yellow	Circular	Raised	Entire
S11a	Opaque red	Circular	Convex	Entire

2.3 Cultivation and Isolation of Cultures

Each of the liquid samples (2.5 mL) was aseptically transferred into a series of 250 mL Erlenmeyer flasks containing 22.5 mL NB medium followed by incubation at 30 °C, 200 rpm for 24 h (Certomat-B, B-Braun). One loopful of bacterial cultures was then transferred onto NA plates and incubated for 24 h at 30 °C for 24 h (Memmert, USA). Serial sub-culturings were carried out until single bacterial colonies were obtained. Similar experimental procedures were repeated for the soil samples using one gram of soil sample as inoculant. Single bacterial colonies were identified via the 16S rRNA gene sequencing analysis carried out by Vivantis Technology Sdn. Bhd., Malaysia. A total of 77 bacterial colonies (Table 2.2) were isolated from solid and liquid samples from the BARC, Johor (45 colonies) and oil refinery wastewater treatment plant at Port Dickson, Negeri Sembilan (32 colonies). Of the 45 colonies isolated from the BARC compound, isolate S8b was chosen for further studies based on its intense yellow-orange coloration when grown in NA and NB mediums (Fig. 2.1a,b). For bacterial colonies isolated from the oil refinery in Port Dickson, 8 were colored. Of these, isolate S1a that displayed gradual color change from grey to dark violet throughout incubation period, was chosen for further study (Fig. 2.1c, d).

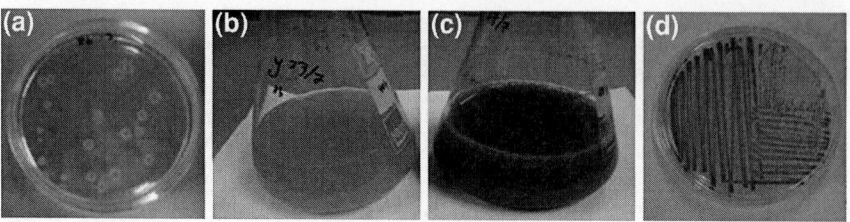

Fig. 2.1 Morphology of isolate S8b on (**a**) NA plate (**b**) NB medium; isolate S1a on (**c**) NB medium (**d**) NA plate

2.4 Characterization of Microorganisms (violet, yellow, red)

The characterization of bacteria was carried out by Vivantis Technologies Sdn. Bhd., Malaysia, using 16S rRNA sequence analysis. From the analysis, the pigment-producing bacteria were identified as follows; yellow–orange as *Chryseobacterium* sp., violet pigment as *Chromobacterium violaceum* (*C. violaceum*) and red pigment as *Serratia marcescens* (*S. marcescens*).

C. violaceum, a Gram-negative bacteria belonging to the Rhizobiaceae family, is a saprophyte found in soil and water in tropical and subtropical areas. In most cases, it is a minor component of the total microflora (Balows et al. 1992). Its colonies are slightly convex, not gelatinous, regular and violets, although irregular variants and non-pigmented colonies can also be found (Sneath 1994). *C. violaceum* has been reported to produce a violet pigment called violacein. Violacein possesses anti-leishimanial (Leon et al. 2001), anti-viral (Andrighetti-Fröhner et al. 2003), antitumoral (Ueda et al. 1994, Melo et al. 2000) and anti-*Mycobacterium tuberculosis* (De Souza et al. 1999) activities. Other properties of *C. violaceum* include the production of cyanide (Michaels and Corpe 1965), the solubilization of gold (Faramarzi et al. 2004), the production of chitinolytic enzymes (Chernin et al. 1998), the synthesis of bioplastics (Steinbüchel et al. 1993) and environmental detoxification (Carepo et al. 2004).

The genus *Chryseobacterium* was created by Vandamme et al. (1994) to accommodate several species formerly classified in the genus *Flavobacterium*, i.e., *Chryseobacterium balustinum*, *C. gleum*, *C. indologenes*, *C. indoltheticum*, *C. meningosepticum* and *C. scophthalmum*. Six species, *Chryseobacterium defluvii* (Kämpfer et al. 2003), *C. joostei* (Hugo et al. 2003), *C. miricola* (Li et al. 2003), *C. daecheongense* (Kim et al. 2005), *C. formosense* (Young et al. 2005) and *C. taichungense* (Shen et al. 2005), have been added to the genus recently.

Chryseobacterium strains produce translucent colonies, shiny with entire edges, but on prolonged incubation the colonies were not visible as single entities probably due to the profuse production of extracellular substances. On NA, it can produce a bright yellow nondiffusible, nonfluorescent flexirubin pigment. It was also reported to have an ability to produce heat stable metalloproteases and protein deamidating enzymes (Venter 1987; Yamaguchi and Yokoe 2000).

2.5 Maintenance of Stock Culture

The LB-glycerol solution was used for long-term storage (months to years) of the bacterial stock culture. The bacteria were first inoculated onto NA plate and incubated at 30 °C for 24 h. The bacterial colonies formed were then transferred into LB-glycerol prior to storage at −20 °C. LB-glycerol was prepared using the following procedure; a mixture of tryptone (10 g), yeast extract (5 g) and NaCl (10 g) were dissolved in 1 L distilled water. pH of the solution was then adjusted to 7.0 using 0.1 M NaOH followed by autoclaving at 121 °C, 121 kPa for 15 min (HVE-50, Hirayama). The bacterial stock cultures were prepared by transferring 2 mL of active culture (from NA plates) into a series of 5 mL Bijou bottles containing 2 mL of glycerol, 25% (v/v) followed by storage at −20 °C prior to use. Another medium used for the maintenance of bacterial stock cultures was 0.1% (w/v) peptone water (Gillis and Logan 2005), which was used specifically for *C. violaceum*. This was due to the inability to revive its growth in broth medium upon storage for 14 days in LB-glycerol. The presence of tryptophan-rich peptic digest of animal tissue in dilute peptone water (MacFaddin 1980) allows the survival of microorganisms up to several years (Gillis and Logan, 2005). Tryptophan is a precursor in the biosynthesis of violacein and its production is essential for pigment production in *C. violaceum* (Vasconcelos et al. 2003). The absence of essential precursors, cofactors and/or accessory proteins in the host organism would disrupt normal functions of various biosynthetic enzymes (August et al. 2000).

2.6 Characterization of Pigments

2.6.1 Extraction of Pigments

The bacterial cells were first grown for 24 h followed by centrifugation at 7500 rpm for 20 min (SIGMA 4 K-15, B.Braun). Both the supernatant and bacterial cell pellets were extracted (7500 rpm, 20 min) using either 95% (v/v) methanol (*S. marcescens* and *C. violaceum*) or 99.5% (v/v) acetone (*Chryseobacterium* sp.)in the ratio of 1: 5 (supernatant) or until the pellet was colorless, i.e., complete pigment extraction has been achieved. The bacterial cell pellet was then discarded while the supernatant was first extracted using ethyl acetate followed by concentration using rotary evaporator (BÜCHI R-210, Switzerland) at 50 °C with chiller temperature set at below 10 °C. The pigment concentration process was carried out until around 1% (v/v) of the initial solvent volume was left in the evaporation flask. The concentrated pigment was then transferred onto glass Petri dishes prior to drying for 3 days at 60 °C.

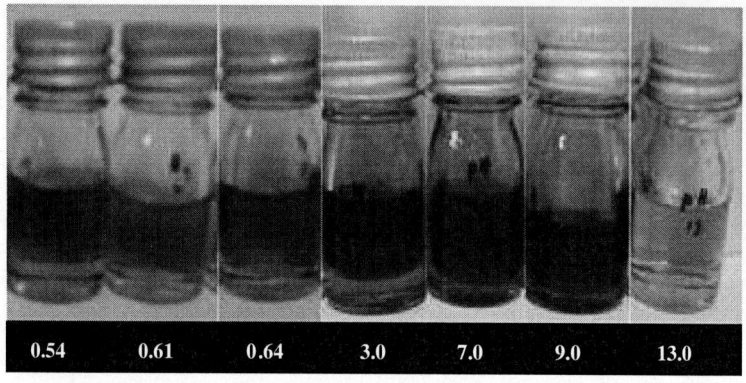

Fig. 2.2 The stability of crude-violet pigment at varying pH

2.6.2 Characterization of Crude-Violet Pigment

UV/vis and FTIR analysis

The purple pigment was tested for stability at various pH ranging from extremely acidic (0.54) to highly basic (13). At extremely low pH values, the purple pigment appeared greenish-blue, bright to dark-blue (3.0–9.0) while in highly basic condition, the pigment was almost decolorized (Fig. 2.2). The decrease in color intensity at high pH values can be attributed to the deprotonation of nitrogen by NaOH from the three conjugated rings at the pigment structure. This resulted in electron conjugations in the ring structure that gives it stability (Konzen et al. 2006). UV/vis analysis (200–800 nm scan; UV 1601PC; Shimadzu) of the crude-violet pigment (Fig. 2.3) showed maximum absorption peaks at 623.67 nm for pH 0.54, 576.61 nm (pH 0.61), 589.81 nm (pH 0.64) and 573.17 nm (pH 3, 7, 9 and 13). Methanol was used as blank.

The primary reason for stronger absorption of the pigment at longer wavelength (visible region) was the electron conjugation effect. A conjugated system requires lower energy for the electronic transition from the Π to Π^* orbitals. The presence of greater number of conjugated bonds resulted in λ_{max} appearing at the longer wavelength region (bathochromic shift) (Mohan 2007). Besides, the auxochromic or chromophoric substitution of five membered heteroaromatics ring present in the crude-violet pigment also causes a bathochromic shift and an increase in the intensity of the bands of the parent molecule. An auxochrome is an auxillary group which interacts with the chromophore resulting in bathochromic shift. The auxochromic group has the ability to provide additional opportunity for charge delocalization, thus providing smaller energy increments for transition to excited states. The charge delocalization by the contributing structures is in the presence and absence of electron donating-NHR group as shown in Fig. 2.4. From the figure, it is clearly shown that the charge delocalization is greatly enhanced by the presence of electron donating-NHR group. The added opportunity for stabilization

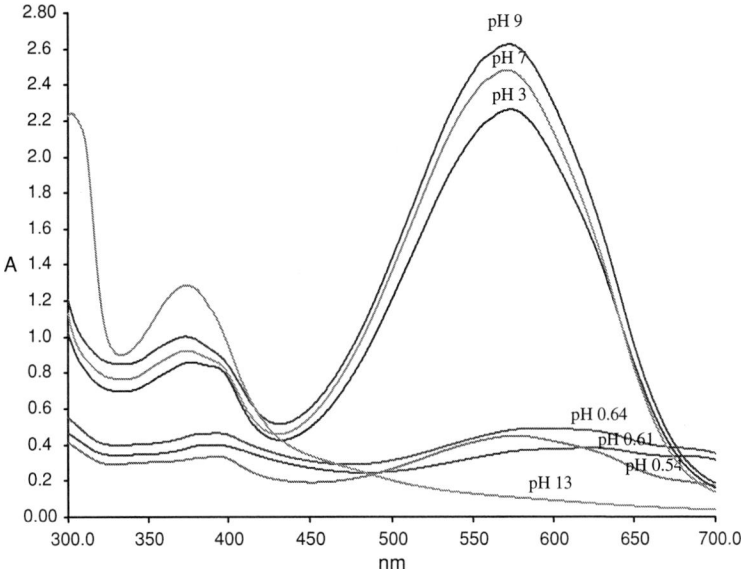

Fig. 2.3 UV/vis spectrum of crude-purple pigment at various pH values

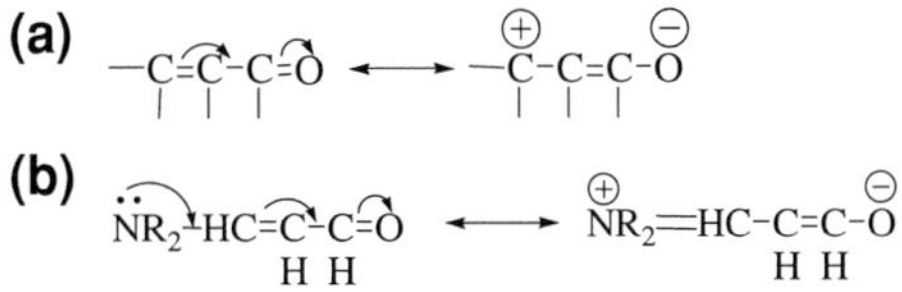

Fig. 2.4 Charge delocalization by the contributing structures in the (**a**) absence and (**b**) presence of electron donating-NHR group (Mohan 2007)

of the Π^* excited state brings the lowest excited state closer to the highest ground state and thus permits a lower energy, i.e., longer wavelength for transition (Mohan 2007).

FTIR analysis of the crude-violet pigment was carried out as follows; The RWS sample (1 mg) was ground with 200 mg of KBr (spectroscopic grade) in a mortar before pressed into 10 mm diameter disks under 6 tons of pressure. FTIR spectra were obtained on a FT–IR 8300, Shidmadzu spectrometer. The analysis conditions used were 16 scans at a resolution of 4 cm^{-1} measured between 400 and 4,000 cm^{-1}. From the spectrum obtained (Fig. 2.5), crude methanolic extract of the violet pigment showed a broad band at 3,700–3,000 cm^{-1} which corresponds to O–H stretching that also overlapped with N–H stretching of a secondary amide at 3256.53 cm^{-1}. Secondary amides are associated through H-bonding to form dimers (cis configuration) or polymers (trans configuration) resulting in the replacement of free N–H stretching band (Fig. 2.6). The weak band at 3256.53 cm^{-1} may be due to an overtone of the band at 1543.93 cm^{-1}

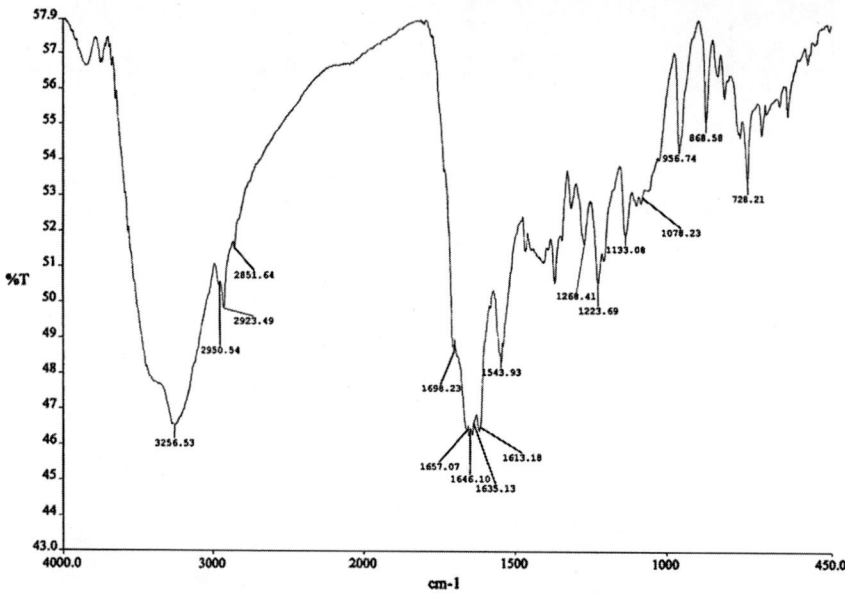

Fig. 2.5 FTIR spectrum for crude methanolic extract of violet pigment. (**a**) cis-dimer (**b**) trans-polymer

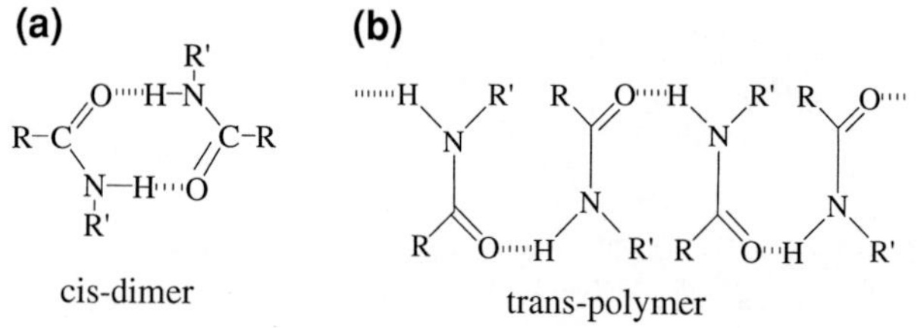

(a)   **(b)**

cis-dimer   trans-polymer

Fig. 2.6 Structure of secondary amides (**a**) ciscoid–ciscoid associations and (**b**) transoid–transoid polymeric association

(trans secondary amide) or combination of C=O stretching and N–H in-plane bending (cis secondary amide) (Mohan 2007).

The C=O stretching of this amide was depicted as doublet at 1657.07 cm^{-1} and 1635.13 cm^{-1}. This doublet is highly characteristic of the amide I band (predominantly C=O, 1635.13 cm^{-1}) and the amide II band (predominantly N–H in-plane bending, 1657.07 cm^{-1}). The frequency of C=O based amide is lower due to the conjugation of the carbonyl group with an aromatic ring. This results in the delocalization of the Π electrons of both unsaturated groups and reduces the double bond character of both bonds, hence reducing the typical carbonyl frequency from 1718 cm^{-1} to 1657.07 cm^{-1} and C=C stretching frequency from 1645 cm^{-1} to 1613.18 cm^{-1}. This situation can be explained further according to Fig. 2.7. In addition, the electron releasing groups (i.e., the amino group), tends to

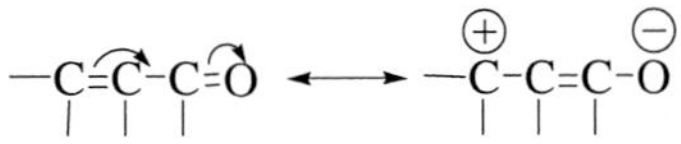

Fig. 2.7 Resonance structure
of carbonyl containing group
in double bond structure
(Mohan 2007)

$$-C{=}C{-}C{=}O \longleftrightarrow \overset{\oplus}{-}C{-}C{=}C{-}\overset{\ominus}{O}$$

Table 2.3 $v_{OH \text{ (free)}}$ cm^{-1}
and v_{C-O} cm^{-1} in alcohols
and phenols

Nature of the hydroxyl compound	$v_{OH \text{ (free)}}$ cm^{-1}	v_{C-O} cm^{-1}
Primary alcohol	3640	1050
Secondary alcohol	3630	1100
Tertiary alcohol	3620	1150
Phenols	3610	1200

favor the polar contributing form by mesomeric effect, thus lowering the force
constant of the C=O bond as well as decreasing the carbonyl-group stretching
frequency. The strong C–O stretching vibration at 1223.69 cm^{-1} is highly characteristic of the C–O stretching vibration of phenol. In fact, further examination of
the C–O stretching vibration is normally used to distinguish between primary,
secondary and tertiary alcohols as shown in Table 2.3.

TLC, NMR and MS analysis

Crude methanolic extract of the violet pigment was spotted on a 0.20 mm pre-
coated silica gel aluminum sheet (Merck Kieselgel 60F$_{254}$) and developed with a
mixture of benzene : acetone (2 : 1) (Duran et al. 1994). The crude pigment
showed a single spot with R$_f$ value of 0.43 (vioacein) when viewed under UV light
at 254 and 356 nm. Upon spraying with the vanillin-sulfuric acid mixture, a small
sport with R$_f$ value of 0.50 appeared which should represent deoxyviolacein, as
suggested by DeMoss and Evans (1959). The vanillin-sulfuric acid mixture was
prepared by mixing vanillin (0.5 g), methanol (80 mL), acetic acid (10 mL) and
concentrated sulfuric acid (5 mL).

Powdered form of the crude extract was analyzed for ^{1}H NMR (400 MHz) and
^{13}C NMR (100 MHz) profiles using Bruker Avance 400 NMR spectrometer.
Chemical shifts (in ppm) were reported relative to tetramethylsilane (TMS) with
deuterated dimethyl sulfoxide (DMSO) as solvent. The ^{1}H NMR spectrum showed
the presence of 13 protons (Fig. 2.8). Three singlet signals were observed at
down-field region which corresponds to indole NH at δ 11.87 ppm, lactam NH
(δ 10.72 ppm) and isatin NH (δ 10.60 ppm). Three aromatic protons in the indole
skeleton which corresponds to the meta-couple and ortho-couple signals were
detected at δ 6.79 ppm (H-7, dd, $J = 2.0$ Hz, 8.8 Hz), δ 7.23 ppm (H-5, d,
$J = 2.0$ Hz) and δ 7.35 ppm (H-8, d, $J = 8.8$ Hz). Other peaks were detected at δ
8.06 ppm (H-2, d, $J = 2.4$ Hz) and δ 9.32 ppm (s, hydroxyl group). Chemical
shifts observed at these lower-field regions were due to hydrogen bonding
occurring between phenolic hydroxyl group with an ortho group (H-5 and H-7)
that exists in the indole skeleton (Lambert and Mazzola 2004). Four protons were
observed in aromatic isatin skeleton. These protons appeared at δ 8.93 ppm (H-19,
d, $J = 7.6$ Hz), δ 6.94 ppm (H-20, t, $J = 7.6$ Hz), δ 7.20 ppm (H-21, t,

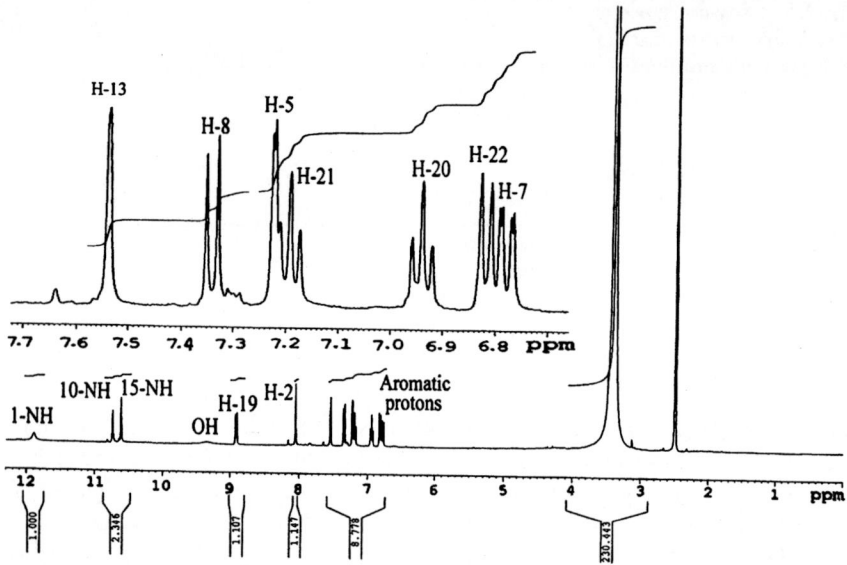

Fig. 2.8 ^{1}H NMR spectrum of violet pigment

Fig. 2.9 The structure of violacein

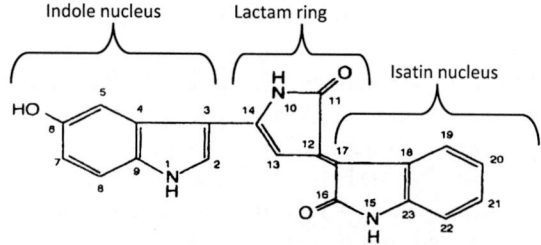

$J = 7.6$ Hz) and δ 6.83 ppm (H-22, d, $J = 7.6$ Hz). A doublet peak at δ 7.55 ppm with $J = 2.0$ Hz was assigned to H-13 in the lactam skeleton. The appearance of low-field signals from indole and pyrrole NH protons can be attributed to the presence of electron-withdrawing group substituents attached to the pyrrole ring. This corresponds to the hydrogen bonding with the solvent resulting in three sharp peaks observed in low-field signal. These three sharp NH signals were the most striking features of violacein as reported by Hoshino et al. (1987) and Yada et al. (2007). The numbering of the structure was cited from Nakamura et al. (2002) (Fig. 2.9).

The ^{13}C NMR spectrum exhibited the presence of 20 carbon atoms (Fig. 2.10). The carbonyl carbons were detected at δ 170.7 ppm (C-16) and δ 172.1 ppm (C-11) while 18 other carbons were observed at δ 97.5 (C-13), 105.1 (C-5), 106.2 (C-3), 109.5 (C-22), 113.6 (C-7), 113.9 (C-8), 119.2 (C-17), 121.3 (C-20), 122.9 (C-18), 126.1 (C-4), 126.8 (C-19), 129.9 (C-2), 130.1 (C-21), 132.1 (C-9), 137.5 (C-12), 142.3 (C-23), 148.1 (C-14) and 153.4 (C-6). The signal at δ 153.4 indicates

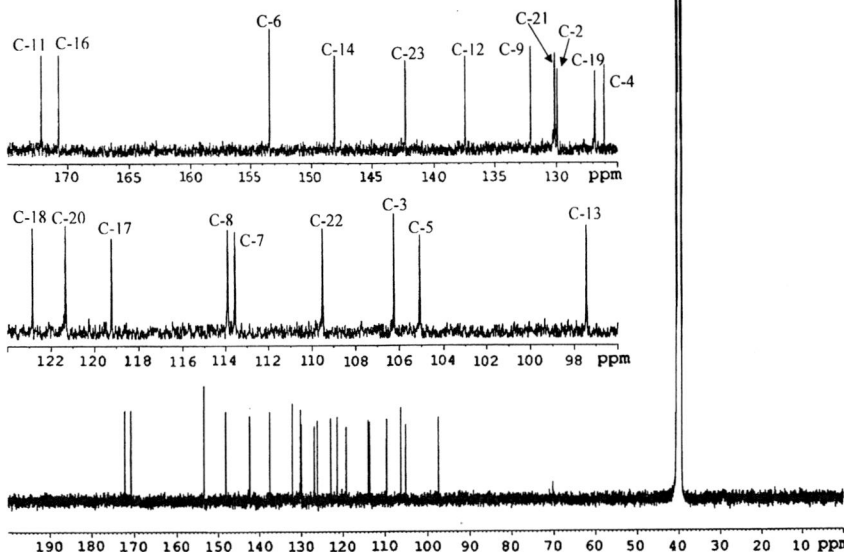

Fig. 2.10 ^{13}C NMR spectrum of violet pigment

that the pigment was violacein, since the signal for corresponding carbon of deoxyviolacein, which lacks hydroxyl residue, appears in the higher magnetic field (Hoshino et al. 1987).

Powdered form of the crude extract (0.5 g) was dissolved in DMSO prior to analysis using the Electron Spray Ionization–Mass Spectrometry (ESI–MS) which was carried out at the National Metrology Laboratory, Standard and Industrial Research Institute of Malaysia (SIRIM), Salak Tinggi, Malaysia. A quasimolecular ion peak was observed at m/z 342.34 [M–H]$^-$ (Fig. 2.11) while the MS/MS spectrum (Fig. 2.12) showed the ion fragments at m/z 298.42, m/z 209.07 and m/z 157.16. These clearly correspond to the molecular formula of $C_{20}H_{13}O_3N_3$, which together with the NMR and FTIR analysis, strongly suggests that the violet methanolic pigment extract is violacein. Similar conclusion was made for violet pigment isolated from other bacteria such as *Janthinobacterium lividium* (Lu et al. 2009), *Duganella* sp. B2 (Wang et al. 2009), *Pseudoalteromonas luteoviolacea* (Yada et al. 2007) and *Alteromonas luteoviolacea*.

Antibacterial activities

The agar disc diffusion method was used to determine the antibacterial activity of the crude-violet pigment as follows (Barja et al. 1989); *Bacillus cereus (B. cereus)*, *Staphylococcus aureus (S. aureus)*, *Pseudomonas aeruginosa (P. aeruginosa)* and *Escherichia coli (E.coli)* were incubated in NB medium either at 30 or 37 °C for 12 h at 200 rpm. Then, 0.1 mL of the culture broth was inoculated onto NA plates containing Whatman filter paper discs (wet strengthened, 0.5 cm in diameter) which were previously impregnated with 50 µL of the crude methanolic extract. Inhibition zones were recorded after overnight incubation at either 30 or 37 °C, respectively.

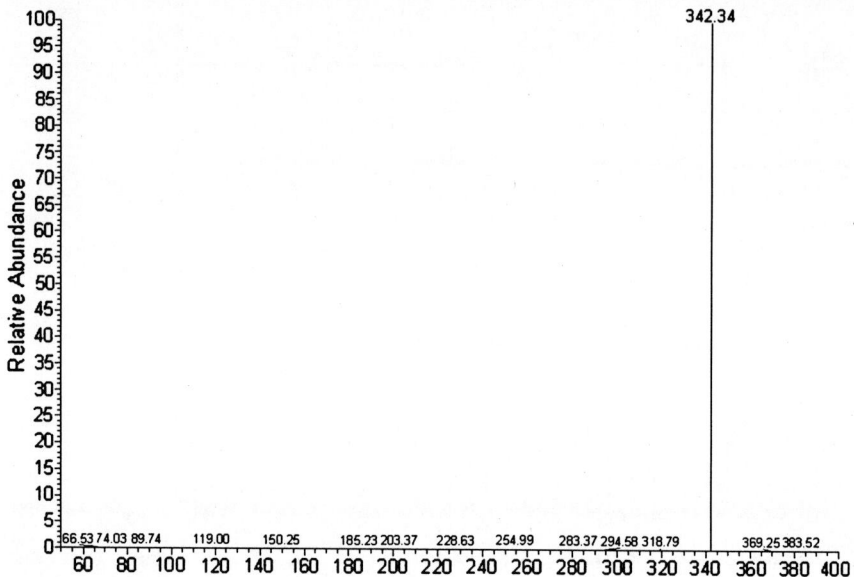

Fig. 2.11 The ESI–MS spectrum of violet pigment

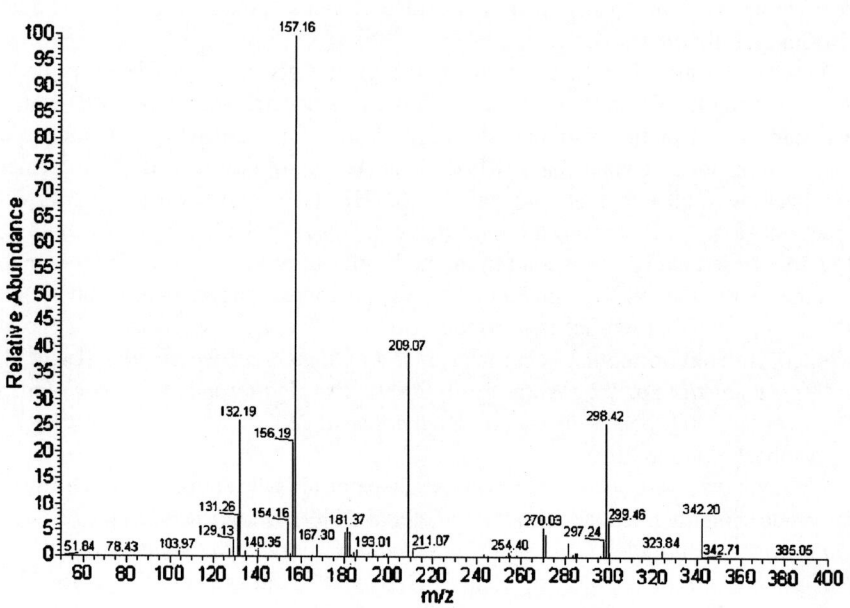

Fig. 2.12 The ESI–MS/MS spectrum of violet pigment ($m/z = 342.34$)

Fig.2.13 Antimicrobial testing of violet pigments on *P. aeruginosa, S. aureus, B. cereus* and *E. coli;* (**a**) – Control (bacteria only); (**b**) – bacteria + methanol (medium suspension for pigment); (**c**) – bacteria + methanol + violet pigment

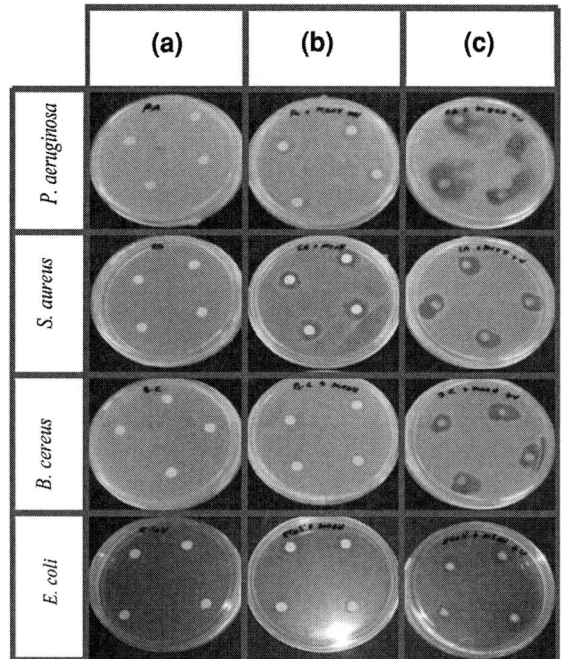

Similar experimental setup consisting of filter paper discs impregnated with 50 μL methanol and filter paper discs only acted as control. The appearance of the inhibition zone on both the Gram-positive (*S. aureus* and *B. cereus*) and the Gram-negative bacteria (*P. aeruginosa* and *E. coli*) indicated that the violet pigment has a broad-spectrum antibacterial property (August et al. 2000) (Fig. 2.13). A slightly larger inhibition zones for the Gram-positive bacteria suggested different anti-bacterial efficiency level of the violet pigment depending on classes of bacteria tested. Among the most important factors influencing the efficacy of antibacterial agents include nature of physical environment and condition of microorganism (Bloomfield 1991), rigidity of the bacterial structure as well as inhibition of the necessary metabolic reaction in a microorganism (Nakamura et al. 2003).

2.6.3 Characterization of Yellow–Orange Pigment

An overnight grown culture of *Chryseobacterium* sp. was centrifuged and the pellet was deposited on a glass slide placed on a white background, then flooded with potassium hydroxide (KOH), 20 % (w/v) and the color shift was observed. The resulting color may then be compared with the initial color of the pellet which acted as a control. After removing excess KOH, an acidic solution was flooded to revert their initial color (Bernardet et al. 2002). The *Flavobacteriaceae* group was reported to produce yellow to orange (or rarely red) pigmentation. As *Chryseobacterium* sp. falls into this group and it produces yellow–orange pigment, a simple test for flexirubin-type pigment was done to diagnose the type of pigment. In this test, a plate containing the *Chryseobacterium* sp. was scrapped and flooded with 20% (w/v) KOH. From the observation, the color of the colony changes from yellow–orange to red–brown. The resulting color was then compared with the bacteria which were not in contact with KOH. After removing the excess KOH, an acidic solution was flooded to revert their initial color. It was suggested that flexirubin reaction was shown in the presence of acid and base condition because it is one diagnostic feature of flexirubin type pigment (Balows et al. 1992). Similar finding was obtained by Kim et al. 2005, where the Flexirubin-type pigment was detected in *Chryseobacterium daecheongense* sp. nov. according to this method. According to Bernardet et al. 2002, since the color shift may pass unnoticed when the KOH solution is poured directly over a thin colony on an agar plate, it is strongly recommended that the test be performed on a small mass of bacterial cells collected with a loop and deposited on a glass slide placed on a white background. Another similar mass of bacteria on which no KOH is poured may be used as a control. The color changed induced by KOH is not absolutely specific for the flexirubin-type of pigment (Fautz and Reichenbach 1980), but it is still helpful when combined with the results of other tests. Natural yellow pigment was tested against common pathogens *Escherichia coli* which is a Gram-negative bacteria and *Bacillus cereus* which is a Gram-positive bacteria. By using the qualitative method, the yellow pigment inhibited the growth of *Bacillus cereus* and *Escherichia coli*. The antibacterial activity of a material depends on the destruction of the physical structure or the inhibition of reaction in a microorganism; it seems that the presence and the level of the antibacterial activity of the pigment varied significantly with the type of microorganism used (Martinko and Madigan 2006). It can be seen from Fig. 2.14 that the yellow pigment has some antimicrobial activity as evidenced from the clearing zones observed in the plates containing *Bacillus cereus* and *Escherichia coli*. To have an idea of the degree of antimicrobial activity, the spread plate method was carried out. A greater percentage kill was observed for the Gram-positive bacteria as compared to the Gram negative bacteria (15 % and 55.61 %, respectively). This could be due to the thinner peptidoglycan layer found in the Gram-negative bacteria making it more susceptible to the pigments.

2.6.4 Characterization of Red Pigment

Extraction and purification

   A modified procedure for the isolation of prodigiosin, a known red pigment produced by *S. marcescens*, was carried out according to Wei et al. (2005); a 24 h old culture broth (1.5 liters), incubated at 25 °C and 200 rpm, was mixed with 4 liters of 95 % (v/v) methanol and mixed vigorously via vortex mixing.

Fig. 2.14 Anti-microbial testing of yellow pigments on *B. cereus* and *E. coli*

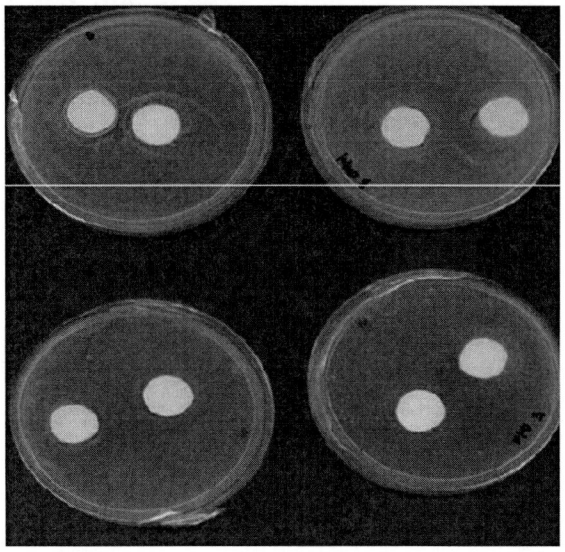

The mixture was then centrifuged at 6854 rpm, 0 °C for 10 min. The resulting supernatant was collected and filtered through a 0.2 μm Whatman filter paper. The filtrate was concentrated using a rotary evaporator (Rotavapor R–200, Buchi) followed by extraction using 5.0 mL chloroform. The chloroform extract was reconcentrated using rotary evaporator until minimal volume was obtained. This minimal volume of chloroform extract was then transferred into a glass petri dish prior to drying in a vacuum dying oven. Purification of the crude product was carried out using column chromatography on a silica gel (230–450 μm, Sigma) column (30 cm height, 2.5 cm internal diameter, 2.9 cm outer diameter). The column content was first washed with hexane. The dried crude product was dissolved in 15 mL of 95% (v/v) methanol before passing through the column. The crude extract - loaded column was eluted with 10.2 M ethyl acetate. Every 8 mL of the orange eluate fractions were collected and examined for the presence of prodigiosin by thin layer chromatography, TLC (silica gel 60 F_{254}, Merck) using ethyl acetate as solvent. Fractions having a single spot were pooled and evaporated in a desiccator to obtain the purified product.

UV–vis, FTIR and NMR analysis

The pigment was analyzed for maximum UV–vis absorbance at pH values of 2, 7 and 9 between 800 and 200 nm, using methanol as blank. FTIR spectroscopy was carried out by mixing the vacuum-dried pigment with finely ground KBr (1:100). Upon pressing under 2,000 kPa, pellet disc obtained was analyzed using a FTIR spectrophotometer (FTIR 8300, Shimadzu) between 4,000 and 400 cm^{-1}. The pigment (5–10 mg) was dissolved in 500 μL of chloroform-d2 ($CDCl_3$) for NMR analysis (Min-jung et al. 2006). UV–vis analysis spectra for the red pigment in methanol showed a maximum peak at 534.76 nm which closely resembles the reported maximum absorbance for prodigiosin, i.e., 535 nm (Song et al. 2000).

Fig. 2.15 Different
coloration of the red pigment
at (**a**) pH 2 (**b**) pH 7 and
(**c**) pH 9

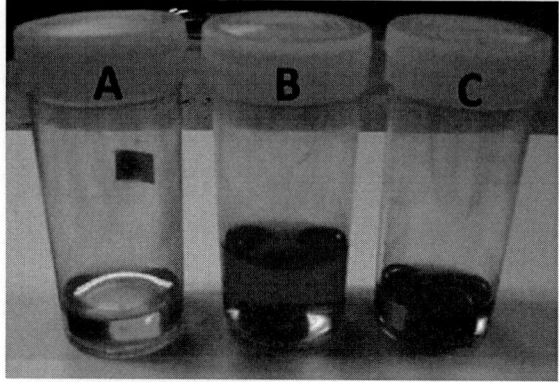

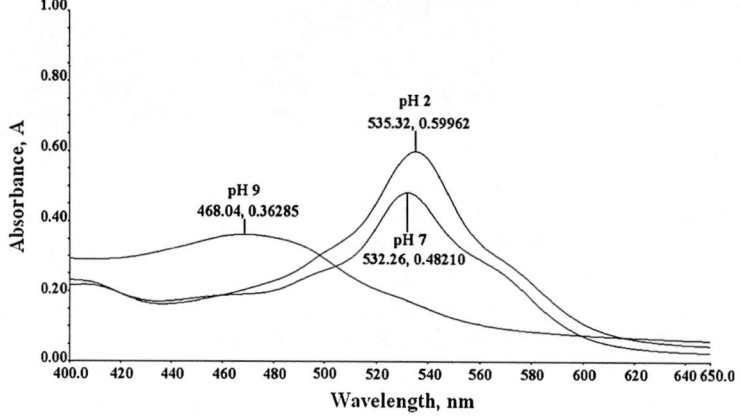

Fig. 2.16 Maximum absorption of the red pigment at pH 2, 7 and 9; pigment dissolved in
methanol

The red-pigment changes to pink at pH 2, orange at pH 9 while retaining its red
feature at pH 7 (Fig. 2.15). Similar observation was previously reported by
Min-jung et al. (2006). At pH 2, maximum absorbance for the pigment was
recorded at λ_{max} 535.32 nm, 532.26 nm at pH 7 and 468.04 nm at pH 9 (Fig. 2.16).

The FTIR spectra of the red pigment showed that it has several degree of
similarity to the spectra of prodigiosin (Song et al. 2000; Rustom et al. 1990).
The summary of the important functional groups found in prodigiosin is as shown
in Table 2.4.

The similar peaks obtained for the red pigment are as shown in Table 2.5. The
main peaks are at 3445.45, 2926.23, 1718.04, 1650.96, 1559.85, 1537.95, 1508.44,
1458.75 and 1072.49 cm^{-1}. From the peaks, it was suggested that the main
functional group for the red pigments are pyrrole, methylene, alkane and alkene.

The significant ^{1}H NMR (CDCl$_3$) chemical shifts obtained for the red pigment
are as follows; 0.0826 mg L^{-1}, 0.5846 mg L^{-1}, 0.6522 mg L^{-1}, 0.8734 mg L^{-1},
1.2637 mg L^{-1}, 1.3756 mg L^{-1}, 1.6071 mg L^{-1}, 2.0478 mg L^{-1}, 2.3585 mg L^{-1},

Table 2.4 Band assignments to the functional groups present in prodigiosin

Wave number (cm^{-1})	Band assignment	Reference
2852	Methylene	Song et al. (2000)
1714	–	
1630	–	
1604	–	
1545	Pyrrole ring structure	
3500	Amide groups	Rustom et al. (1990)
2800	Methylene	
1500, 1565	Pyrrole ring structure	

Table 2.5 FTIR band assignments for red pigment

Wave number (cm^{-1})	Band assignments	Wave number (cm^{-1})	Band assignments
3445.45	Aliphatic alcohols (s) Primary amines (m), amide (m)	1537.95	C=C stretching vibrations
2926.23	C–H stretching (s) Methylene	1508.44	Secondary amines (w) Azo compounds (w) Aromatic (m)
1718.04	C=O stretching vibrations Ester (s) Aryl aldehydes (s) Saturated ketone (s) Saturated carboxylic acid (s)	1458.75	C–H bending vibrations (m) Ethylenediamine complex (m) Azo compounds (w)
1650.96	C=C Sretching Vibrations Aliphatic amines (m) Enol (s)	1072.49	C–H out-of-plane bending vibrations Aliphatic amine (w-m)
1559.85	Secondary amines (w)		

m moderately strong, *s* strong, *w-m* weak to moderately strong, *w* weak

3.6781 mg L^{-1}, 3.8127 mg L^{-1}, 4.1521 mg L^{-1} and 5.3590 mg L^{-1}. Shift for the solvent was at 7.2734 mg L^{-1}. The main functional groups observed were: saturated alkane, amine, methyl (primary), methyl (secondary), methyl (tertiary), esters, acetylenic, ethers, alcohols and vinylic. It was observed that the signals of the vinylic functional group (C=C) were very weak in the NMR spectra.

References

Andrighetti-Fröhner CR, Antonio RV, Creczynski-Pasa TV, Barardi CRM, Simões CMO (2003) Cytotoxicity and potential antiviral evaluation of violacein produced by *Chromobacterium violaceum* Memórias do Instituto Oswaldo Cruz 98:843–848

August PR, Grossman TH, Minor C, Draper MP, MacNeil IA, Pemberton JM, Call KM, Holt D, Osburne MS (2000) Sequence analysis and functional characterization of the violacein biosynthetic pathway from *Chromobacterium violaceum*. JMMB 2(4):513–519

Balows A, Truper HG, Dworkin M, Harder W, Schleifer KH (1992) The prokaryotes, 2nd edn. Springler-Verlag, Berlin

Barja JL, Lemos ML, Toranzo EA (1989) Purification and characterization of an antibacterial substance produced by a parine *Alteromonas* Species. Antimicrob Agents Chemother 33(10):1674–1679

Bernardet JF, Nakagawa Y, Holmes B (2002) Proposed minimal standards for describing new taxa of the family *Flavobacteriaceae* and emended description of the Family. Int J Syst Evol Microbiol 52(3):1049–1070

Carepo MSP, Azevedo JSN, Porto JIR, Bentes-Souza AR, Batista JS, Silva ALC, Schneider MPC (2004) Identification of *Chromobacterium violaceum* genes with potential biotechnological application in environmental detoxification. Genet Mol Res 3:181–194

Chernin LS, Winson MK, Thompson JM, Haran S, Bycroft BW, Chet I, Williams P, Stewart GSAB (1998) Chitinolytic activity in *Chromobacterium violaceum*: substrate analysis and regulation by quorum sensing. J Bacteriol 180:4435–4441

De Souza AO, Aily DCG, Sato DN, Duran N (1999) Atividade da violaceina in vitro sobre o *Mycobacterium turbeculosis* H37RA, *Rev.* Inst. Adolfo Lutz. 58:59–62

DeMoss RD, Evans NR (1959) Physiological aspects of violacein biosynthesis in nonproliferating cells. J Bacteriol 78:583–586

Eaton AD, Franson MAH (2005) Standard methods for the examination of water and wastewater, 21st edn. American Public Health Association, New York

Faramarzi MA, Stagars M, Pensini E, Krebs W, Brandl H (2004) Metal solubilization from metal-Containing solid materials by cyanogenic *Chromobacterium violaceum*. J Biotechnol 113:321–326

Fautz E, Reichenbach H (1980) A simple test for flexirubin-type pigments. FEMS Microbiol Lett 8:87–91

Gillis M and Logan N A (2005) Genus IV. *Chromobacterium Bergonzini* 1881, 153AL. In: Brenner DJ, Krieg N ,Staley JT, Garrity GM(eds.). Bergey's manual of systematic bacteriology, 2nd edn, vol 2, part C. Springer,New York, pp 824–827

Hoshino T, Kondo T, Uchiyama T, Ogasawara N (1987) Biosynthesis of violacein: A novel rearrangement in tryptophan metabolism with 1, 2-shift of the indole Ring. Agr Chem Biotechno 51:965–968

Hugo CJ, Segers P, Hoste B, Vancanneyt M, Kersters K (2003) *Chryseobacterium joostei* sp. nov., isolated from the dairy environment. Int J Syst Evol Microbiol 53:771–777

Kämpfer P, Dreyer U, Neef A, Dott W, Busse H-J (2003) *Chryseobacterium defluvii* sp. nov., isolated from wastewater. Int J Syst Evol Microbiol 53:93–97

Kim KK, Bae H-S, Schumann P, Lee S-T (2005) *Chryseobacterium daecheongense* sp. nov., isolated from freshwater lake sediment. Int J Syst Evol Microbiol 55:133–138

Konzen M, Marco DD, Cordova CAS, Vieira TO, Antonio RV, Creczynski-Pasa TB (2006) Antioxidant properties of violacein: ossible relation on its biological function. J Bioorg Med Chem 14:8307–8313

Lambert JB, Mazzola EP (2004) Nuclear magnetic resonance spectroscopy. An introduction to Principles, applications, and experimental methods. Pearson Education, USA, pp 75–76

Leon LL, Miranda CC, De Souza AO, Durán N (2001) Antileishmanial activity of the violacein extracted from *Chromobacterium violaceum*. J Antimicrob Chemother 48:449

Li Y, Kawamura Y, Fujiwara N, Naka T, Liu H, Huang X, Kobayashi K, Ezaki T (2003) *Chryseobacterium miricola* sp. nov., a novel species isolated from condensation water of space station Mir. Syst Appl Microbiol 26:523–528

Lu Y, Wang L, Xue Y, Zhang C, Xing XH, Lou K, Zhang Z, Li Y, Zhang G, Bi J, Su Z (2009) Production of violet pigment by a newly isolated psychrotrophic bacterium from a glacier in Xnjiang, China. Biochem Eng J 43:135–141

MacFaddin J (1980) Biochemical tests for identification of medical bacteria, 2nd edn. Williams and Wilkins, Baltimore

Martinko JM, Madigan MT (2006) Brock: biology of microorganism, 11th edn. Pearson Education International, USA

Melo PS, Maria SS, Vidal BC, Haun M, Durán N (2000) In Vitro Cell Dev Biol Anim 36: 539–543

Michaels R, Corpe WA (1965) Cyanide formation by *Chromobacterium violaceum*. J Bacteriol 89:106–112

Min-jung S, Jungdon B, Due-Sil L, Chang-Ho K, Jun-Seok K, Seung-Wook K, Suk-In H (2006) Purification and characterization of prodigiosin produced by integrated bioreactor from *Serratia* sp. KH-95. JBB 101:157-161.

Mohan J (2007) Organic spectroscopy. Principles and Applications. Alpha Science International Ltd., U.K

Nakamura Y, Sawada T, Morita Y, Tamiya E (2002) Isolation of a psychrotrophic bacterium from the organic residue of a water tank keeping rainbow trout and antibacterial effect of violet pigment produced from the strain. Biochem Eng J 12:73–80

Nakamura Y, Asada C, Sawada T (2003) Production of antibacterial violet pigment by psychrotropic bacterium RT102 Strain. Biotechnol Bioprocess Eng 8:37–40

Singh R, Jain A, Panwar S, Gupta D, Khare SK (2005) Antimicrobial activity of some natural dyes. Dyes and Pigments 66: 99-102

Rustom SM, Valiollah H, Alka MP, Prafulla JD (1990) Isolation and characterization of *Serratia marcescens* mutants defective in prodigiosin biosynthesis.Curr Microbio 20(2):95–103

Shen F-T, Kämpfer P, Young C–C, Lai W-A, Arun AB (2005) *Chryseobacterium taichungense* sp. nov., isolated from contaminated soil. Int J Syst Evol Microbiol 55:1301–1304

Sneath PH (1994) *Chromobacterium Bergonzini* 1881. In: Gibbons RE(ed),Bergey's manual of determinative bacteriology, 8th edn. Williams and Wilkins, Baltimore,p 354

Song C, Makoto S, Osamu J, Shinji O, Yasunori N, Akihiro Y (2000) High production of prodigiosin by *Serratia marcescens* grown on ethanol. Biotechnol Lett 22(22):1761–1765

Steinbüchel A, Debzi EM, Marchessault RH, Timm A (1993) Synthesis and production of poly (3-hydroxyvaleric acid) homopolyester by *Chromobacterium violaceum*. Appl Microbiol Biotechnol 39:443–449

Ueda H, Nakajima H, Hori Y, Goto T , Okuhara M (1994) FR901228, a novel antitumor bicyclic depsipeptide produced by *Chromobacterium violaceum* n°. 968. I. taxonomy, fermentation, isolation, physico-chemical and biological properties. J Antibiot (Tokyo) 47: 301-310

Vandamme P, Bernardet J-F, Segers P, Kersters K, Holmes B (1994) New perspectives in the classification of the flavobacteria: description of *Chryseobacterium* gen. nov., *Bergeyella* gen. nov., and *Empedobacter* nom. rev. Int J Syst Bacteriol 44:827–831

Vasconcelos ATR, Almeida DF, Hungria M, Guimarães CT, Antônio RV, Almeida FC, Almeida LGP, Almeida R, Alves-Gomes JA, Andrade EM, Araripe J, Araujo MFF, Astolfi-Filho S, Azevedo V, Baptista AJ, Bataus LAM, Baptista JS, Belo A, van den Berg C, Bogo M, Bonatto S, Bordignon J, Brigido MM, Brito CA, Brocchi M, Burity HA, Camargo AA, Cardoso DDP, Carneiro NP, Carraro DM, Carvalho CMB, Cascardo JCM, Cavada BS, Chueire LMO, Creczynski-Pasa TB, Cunha Junior NC, Fagundes N, Falcão CL, Fantinatti F, Farias IP, Felipe MSS, Ferrari LP, Ferro JA, Ferro MIT, Franco GR, Freitas NSA, Furlan LR, Gazzinelli RT, Gomes EA, Gonçalves PR, Grangeiro TB, Grattaplaglia D, Grisard EC, Hanna ES, Jardim SN, Laurino J, Leoi LCT, Lima LFA, Loureiro MF, Lyra MCCP, Madeira HMF, Manfio GP, Maranhão AQ, Martins WS, Mauro SMZ, Medeiros SRB, Meissner RV, Moreira MAM, Nascimento FF, Nicolas MF, Oliveria JG, Oliveira SC, Paixão RFC, Parente JA, Pedrosa FO, Pena SDJ, Pereira JO, Pereira M, Pinto LSRC, Pinto LS, Porto JIR, Potrich DP, Ramalho Neto CE, Reis AMM, Rigo LU, Rondinelli E, Santos EBP, Santos FR, Schneider MPC, Seuanez HN, Silva AMR, Silva ALC, Silva DW, Silva R, Simões IC, Simon D, Soares CMA, Soares RBA, Souza EM, Souza KRL, Souza RC, Steffens MBR, Steindel M, Teixeira SR, Urmenyi T, Vettore A, Wassem R, Zaha A, Simpson AJG (2003) The complete genome of *Chromobacterium violaceum* reveals remarkable and exploitable bacterial adaptability. Proc Natl Acad Sci USA 100:11660–11665

Venter H (1987) Purification and characterization of a heat stable metalloprotease from a *Chryseobacterium* of dairy origin. MSc thesis. University of Orange Free State, Bloemfontein, South Africa.

Wang H, Jiang P, Lu Y, Ruan Z, Jiang R, Xing XH, Lou K, Wei D (2009) Optimization of culture conditions for violacein production by a new strain of *Duganella* sp. B2. Biochem Eng J 44:119–124

Wei YH, Yu WJ, Chen WC (2005) Enhanced undecylprodigiosin production from *Serratia marcescens* SS-1 by medium formulation and amino-acid supplementation. J Biosci and Bioeng 100:466–471

Yada S, Wang Y, Zou Y, Nagasaki K, Hosokawa K, Osaka I, Arakawa R, Enomoto K (2007) Isolation and characterization of two groups of novel marine bacteria producing violacein. Mar Biotechnol 10:128–132

Yamaguchi S, Yokoe M (2000) A novel protein-deamidating enzyme from *Chryseobacterium proteolyticum* sp. nov., a newly isolated bacterium from soil. Appl Environ Microbiol 66:3337–3343

Young CC, Kämpfer P, Shen FT, Lai WA, Arun AB (2005) *Chryseobacterium formosense* sp. nov., isolated from the rhizosphere of *Lactuca Sativa* L. (garden lettuce). Int J Syst Evol Microbiol 55:423–426

Chapter 3
Optimization of Pigment Production: Case of *Chromobacterium violaceum* and *Serratia marcescens*

Abstract Due to the high cost of the technology currently used for pigment production on an industrial scale, there is a need to develop a low cost process such as the use of agricultural-waste residues as growth medium, instead of the typical expensive synthetic medium. The use of these nutrient-rich agricultural wastes, which is renewable, abundant and easily available, even offers the potential for the production of value-added products such as specialty chemicals, biofuels and bioplastics. It also provides an ingenious way of protecting the environment by reducing the amount of waste to be treated, hence reducing the threat of environmental contamination. However, the pigment-producing bacteria needs to be adapted to grow in these agricultural-waste residues taking into consideration important growth parameters such as temperature, growth medium and light. Temperature is an important factor as it influences metabolic activities and microbial growth, light may influence the production of photosensitive pigment (directly affecting pigment intensity) while knowledge on the bacterial ability to grow either in solid or liquid growth medium is essential to ensure most of the available agricultural-waste residues can be effectively utilized as growth medium.

Keywords Pigment · Bacteria · Production · Temperature · Growth · Medium · Light

3.1 Culture Preparation

Single colony of *C. violaceum* on nutrient agar was used in all experiments. The strain was cultivated in nutrient broth for 16 h before inoculation. Unless otherwise stated, the cultivation was carried out in 250 mL Erlenmeyer flasks shaken at 200 rpm, 30 °C for 24 h. L-tryptophan stock solution was prepared by dissolving 1 g of L-tryptophan powder in 0.1 M NaOH and topped up with

W. A. Ahmad et al., *Application of Bacterial Pigments as Colorant*,
SpringerBriefs in Molecular Science, DOI: 10.1007/978-3-642-24520-6_3,
© The Author(s) 2012

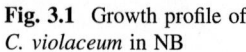
Fig. 3.1 Growth profile of
C. violaceum in NB

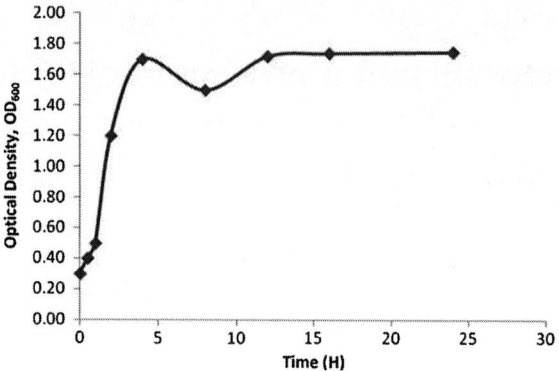

distilled water in a 1 L volumetric flask. The solution was then neutralized using 0.1 M HCl and sterilized using a hydrophobic-edge 0.45 μm Whatman filter paper.

Growth Profile of C. violaceum: A 12 h-grown culture of bacteria (20 mL) was inoculated into a 2 L Erlenmeyer flask containing 180 mL of NB followed by incubation at 200 rpm, 30 °C for 24 h. Culture turbidity (OD_{600}) was measured at regular intervals using a spectrophotometer (Genesys 20, ThermoSpectronic). Pigment production by *C. violaceum* was also recorded. Similar experimental setup as above minus the bacterial cells acted as a control. Growth of *C. violaceum* in NB was monitored for 24 h (Fig. 3.1). The bacteria showed a typical growth curve with distinct log, stationary and death phase. Lag phase was not observed as exponential growing culture was used to inoculate the medium causing cells to commence growth immediately. Optical density reading decreased slightly but increased and remained constant till 24 h. This might be due to the formation of precipitates after 5 h which might interfere with the optical density readings.

It is interesting to note that pigment production by bacteria was observed 4 h after growth i.e., at the late exponential phase. This seems to suggest that the pigments produced by bacteria are secondary metabolites as the color of the pigment was produced after the active stage of growth (Fig. 3.2) (Lu et al. 2009). Intensity of the violacein produced by *C. violaceum* was maximum at 4 h of growth and decreased with time due to precipitation of the pigments. As violacein is reported to be poorly water soluble (De Azevedo et al. 2000) this might explain the decreased intensity of the violet color with time.

Growth Profile of Serratia marcescens: *Serratia marcescens* (*S. marcescens*) was isolated from an oxidation pond at Universiti Teknologi Malaysia, Johor, Malaysia. *S. marcescens* was cultivated in NB (8 g L^{-1}, Merck) and shaken at 200 rpm (Certomat, B. Braun at 25 °C. It was identified via the 16S rRNA gene sequencing analysis carried out by Vivantis Technologies Sdn. Bhd., Malaysia where a 99.0% similarity with *S. marcescens* (AB244453, AB244433, AB244291, EF415649, AB270613, AY566180) was obtained from the reverse nucleotide sequence of 700 bp (Fig. 3.3). The nucleotide sequence was deposited in GenBank where it was given the accession number EU555434.

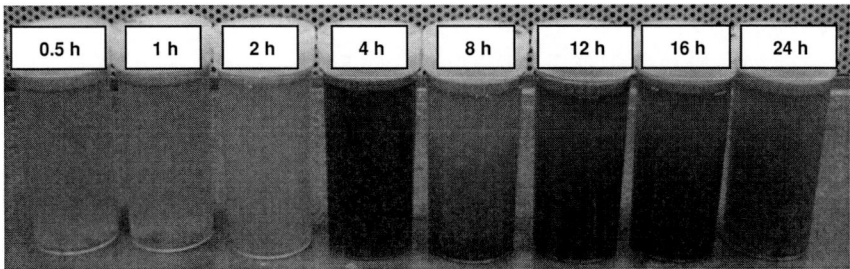

Fig. 3.2 Effect of time on pigment production on *C. violaceum*

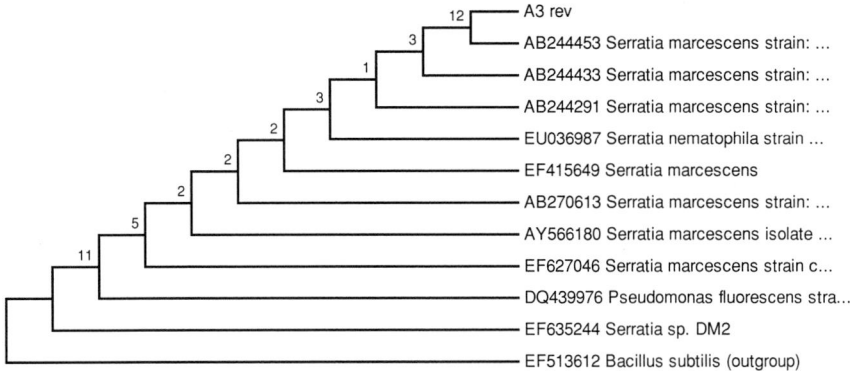

Fig. 3.3 The phylogenetic tree showed the interrelationship between the locally isolated *S. marcescens* (reverse sequence 700 bp, labeled as A3 rev) and top 10 Blast hits from NCBI

Preparation of Nutrient broth (NB) was carried out by dissolving 8 g NB (Merck) in 1 L of deionized water and sterilized by autoclaving at 121 °C, 101.3 kPa for 15 min. Nutrient agar (NA) was prepared by dissolving 20 g of nutrient agar (MERCK, Germany) in 1 L of deionized water. The medium was sterilized by autoclaving at 121 °C, 101.3 kPa for 15 min. The molten agar was cooled to about 50 °C before being poured into sterile Petri dishes. The agar was allowed to harden and then incubated for 24 h at 30 °C to ensure that the medium was free from contamination. In this study, liquid pineapple waste was used as a cheap alternative medium for bacterial growth. The liquid pineapple waste (LPW) was obtained from the downstream process at Lee Pineapple Manufacturing Industry, Tampoi, Malaysia. Prior to use, the LPW was filtered using muslin cloth followed by centrifugation (SIGMA 4 K-15, B. Braun) at 7,000 rpm for 5 min to remove solid particulates. To kill the endogenous microorganisms in LPW, ethanol 5% (v/v) (HmbG® Chemicals) was added as disinfectant. Finally the LPW was neutralized using NaOH (QRëC) 1 M (Salmiah 2006). The brown sugar (BS) stock solution was prepared by dissolving 200 g of BS in 1 L of deionized water. The solution was filtered (Whatman, UK) to remove insoluble materials

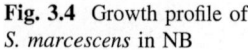

Fig. 3.4 Growth profile of
S. marcescens in NB

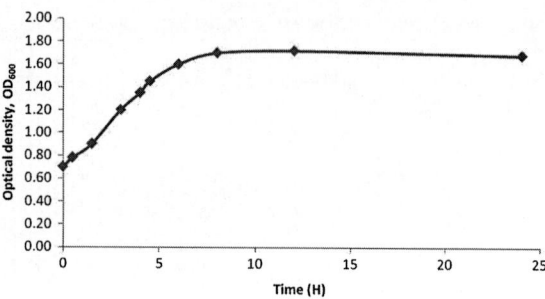

prior to autoclaving at 121 °C, 101.3 kPa for 15 min (HVE-50, Hirayama). Figure 3.4 shows the growth profile of *S. marcescens* in NB.

For this growth profile, lag phase was not visible as exponential growing culture was used to inoculate the medium causing cells to commence growth immediately. The log phase lasted for about 8 h after inoculation into fresh medium. After 8 h the growth was slow and the population reached the stationary phase. This might be due to the depletion of nutrients present in the growth medium (Nancy and Gauthier 1979). During this phase, the production of red pigments was eminent and the color intensity increased with time. This could suggest that the pigment was a secondary metabolite product because the color of the pigments was produced after the active stage of growth (Nancy and Gauthier 1979). It has been reported that prodigiosin (red pigment) is a multifaceted secondary metabolite which is produced by *S. marcescens* and other bacteria such as *Pseudomonas magneslorubra* and *Vibrio psychroerythrous* (Anuradha et al. 2004).

As shown in Fig. 3.5, the bacterium starts to produce color after 8 h of growth and the intensity of color increases with time. It was reported that the induction of pigment formation was seen after 8 h of inoculation whereas the maximum production of pigments by *S. marcescens* was seen from 15 to 36 h (Anuradha et al. 2004), quite similar to our findings.

3.2 Effect of Temperature

C. violaceum: Active cultures of bacteria (10 mL) were inoculated into a series of 1 L Erlenmeyer flasks containing 90 mL NB. The mixtures were shaken at 200 rpm for 24 h at 25, 30 and 37 °C. These temperatures were selected so that large scale production of pigments can be carried out at room temperature, hence saving cost on electricity. At the end of the bacterial growth cycle, the cell dry weight and pigment production were determined. The cell dry weight was determined as follows; the culture broth was centrifuged at 7,000 rpm for 20 min where the pellet obtained was suspended in a small volume of distilled water. Cell suspension (5 mL) was then filtered using a hydrophobic-edge 0.45 μm Whatman filter paper prior to drying at 50 °C for 48 h. The violet pigment produced by

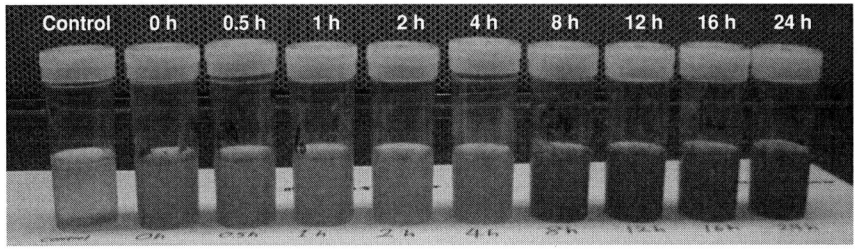

Fig. 3.5 Time-scale representation on the production of red-pigment by *S. marcescens* EU555434 in NB

Fig. 3.6 Growth profile of *C. violaceum* in NB at 25, 30 and 37 °C

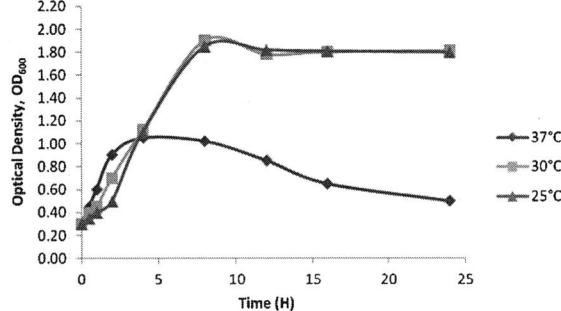

C. violaceum grown at 25, 30 and 37 °C were extracted with 6 mL of ethyl acetate. The three extracts were then analyzed using UV–vis spectrophotometer (Perkin Elmer). *C. violaceum* showed good growth and intense pigment production at 25 and 30 °C as compared to 37 °C (Fig. 3.6).

Temperature is an important factor as it influences metabolic activities and microbial growth. At 37 °C, the metabolic activities of the bacteria decreases, thus affecting cell growth and pigment production by the bacteria. However, *C. violaceum* showed the ability to produce pigment after 4 h of growth at 30 °C. At 25 °C, pigment production was visible only after 8 h of growth. UV–vis analysis of the pigment produced by bacteria showed high absorbance at 30 °C compared to 25 and 37 °C (Fig. 3.7). This was because the bacterium was at its optimum temperature for growth, hence higher amount of cells were produced. This was substantiated by the higher cell dry weight obtained at 30 °C (1.03 mg mL^{-1}) compared to 0.97 mg mL^{-1} at 25 °C and 0.48 mg m L^{-1} at 37 °C .

S. marcescens: Active culture of *S. marcescens* was prepared by inoculating a loopful of bacterial cells from NA plate into 25 mL of NB. The culture was incubated at 25 °C with agitation at 200 rpm using an orbital shaker (Certomat $^{®}$R, B. Braun) for 12 h. The optimization of temperature was evaluated in both NB and one type of agricultural waste i.e., LPW. This was carried out by cultivating 20 mL of the active culture into 180 mL of either medium in 1 L of Erlenmeyer flask (in triplicates). The cultures were incubated at either 25, 30 or 37 °C and shaken at 200 rpm for 24 h. In NB, the medium turned dark red (after 8 h) at

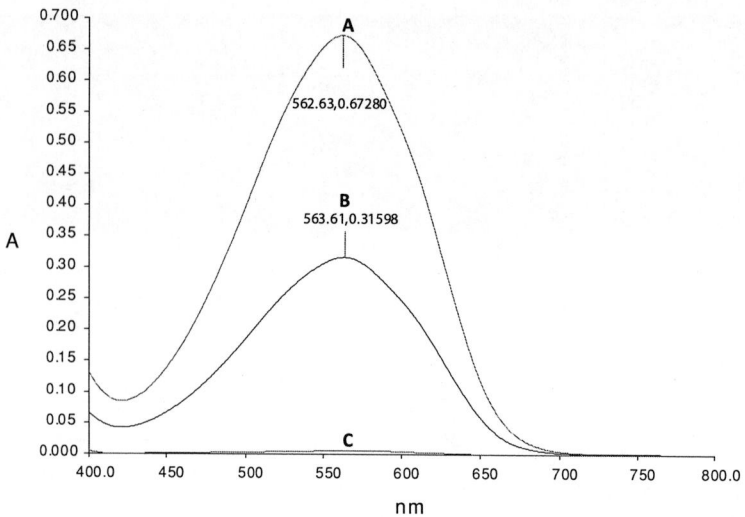

Fig. 3.7 UV-vis spectra for violet pigment produced by *C. violaceum* when grown at (**a**) 30 °C (**b**) 25 °C and (**c**) 37 °C

Fig. 3.8 Intensity of red pigment grown at (**a**) 25 °C (**b**) 30 °C and (**c**) 37 °C

25 °C, pink at 30 °C while no color change was observed at 37 °C. While in LPW, pigment production was observed at all temperatures evaluated with maximum pigment production obtained at 25 °C based on the most intense red coloration compared to 30 or 37 °C (Fig. 3.8). Thus, it can be said that temperature directly affects pigment production by *S. marcescens* based on its role in determining optimum condition required in the regulation of enzyme for bacteria. At higher temperature, certain enzymes required to synthesize the red pigment might be denatured. It has been reported that maximum yield for pigment production by *S. marcescens* was obtained at 28 and 30 °C in NB after 36 h of growth. However, pigment production decreased significantly when incubated at 37 °C (Anuradha et al. 2004).

Fig. 3.9 Different types of
agricultural wastes used
during the study on the effect
of growth media on pigment
production. (**a**) solid
pineapple waste (SPW),
(**b**) brown sugar (BS),
(**c**) liquid pineapple waste
(LPW) and (**d**) molasses (M)

3.3 Effect of Growth Media

C. violaceum: The production of crude violacein in different growth media was
evaluated as follows; 200 mL of either Luria–Bertani (LB) [tryptone 10 g/L, yeast
extract 5 g/L, NaCl 10 g/L], nutrient broth (NB) [peptone 10 g/L, NaCl 5 g/L,
yeast extract 3 g/L], tryptic soy broth (TSB) [peptone from casein 17 g/L, peptone
from soymeal 3 g/L, D-(+) glucose 2.5 g/L, NaCl 5 g/L, Na_2HPO_4 2.5 g/L] or
peptone glycerol broth (PGB) [meat extract 10 g/L, peptone 10 g/L, glycerol 10%
(v/v)] were transferred into a series of 1 L Erlenmeyer flasks. The pH of all the
above media was maintained at 7.0. The various media were autoclaved at 121 °C
for 15 min. Fermentations were carried out using an 16 h old bacterial inoculum
(10% v/v) which was then transferred into the respective flasks followed by
incubation at 200 rpm, 30 °C for 24 h. Samples were withdrawn periodically for
measurements of cell concentration (OD_{600}), pH and concentration of violacein.

Results obtained indicated that maximum cell turbidity (OD_{600}) of 1.83 was
achieved in LB followed by TSB (OD_{600} of 1.81), NB (1.66) and PGB (1.15).
Bacterial growth in all mediums showed an increasing pH profile except PGB,
which was probably due to initiation of pigment production. Maximum production
of violacein was observed in NB (0.15 g L^{-1}) followed by LB (0.13 g L^{-1}) and
TSB (0.05 g L^{-1}). No pigment production was observed in PGB medium which
somewhat did not correlate with its relatively good growth. This may imply that
violacein was not required for growth and survival of *C. violaceum* (Sivendra and
Lo 1975; Durán and Faljoni-Alario 1980).

Another evaluation on the production of pigment by *C. violaceum* in different
growth media was carried out using both the solid and liquid agricultural wastes as
substrates namely LPW, solid pineapple waste (SPW), molasses (M), brown sugar
(BS) and sugarcane bagasse (SCB) (Fig. 3.9). LPW and SPW were obtained from
the waste collection pond of one pineapple processing premise in Tampoi,
Malaysia, BS and M were purchased from the local sundry shop while SCB was

Fig. 3.10 Production of violet pigment by *C. violaceum* when grown at different concentrations of molasses (M), brown sugar (BS), sugarcane bagasse (SCB) and solid pineapple waste (SPW) at 30 °C. (**a**) 0.1% (v/v) M, (**b**) 0.5% (v/v) M, (**c**) 1.0% (v/v) M, (**d**) 1% (v/v) BS, (**e**) 5% (v/v) BS, (**f**) 10% (v/v) BS, (**g**) 1 g SCB (SCB1), (**h**) 3 g SCB (SCB3), (**i**) 5 g SCB (SCB5), (**j**) 1% (v/v) SPW, (**k**) 5% (v/v) SPW and (**l**) 10% (v/v) SPW; flasks (**g–h**) were supplemented with 10% (v/v) L-tryptophan

obtained from local sugarcane processing company. SCB was first dried at 30 °C and cut into a length of between 1.2 and 1.5 cm. The BS stock solution (40 g L^{-1}) was filtered using Whatman filter paper (No. 1) while the molasses stock solution (10% v/v) was adjusted to pH 7.0 using 0.1 M NaOH. Prior to use, both solutions were autoclaved at 105 °C, 101.3 kPa for 15 min. For the solid substrates i.e., SPW and SCB, 1–10% (w/v) of the substrates were transferred into a series of 1 L Erlenmeyer flasks containing 100 mL of deionized water. The mixtures were then added with 100 mg L^{-1} of L-tryptophan followed by pH adjustment to 7.0. L-tryptophan was used as a precursor for pigment production in *C. violaceum* (Vasconcelos et al. 2003). Then, 10% (v/v) active cultures of *C. violaceum* were inoculated into the flasks prior to a 24 h incubation period at 200 rpm and 30 °C. Similar experimental setup was used for the liquid substrates where various concentrations of brown sugar, 1–10% (v/v) and molasses, 0.1–1.0% (v/v) were used. All experiments were complimented with control sets i.e., in the absence of *C. violaceum*.

C. violaceum showed the capability to produce pigment in both liquid and solid substrates (Fig. 3.10) with highest pigment production of 0.82 g L^{-1} achieved in a mixture consisting of 3 g SCB and 10% (v/v) L-tryptophan (labeled as SCB3 medium). Lowest pigment production i.e., 0.19 g L^{-1} occurred when the bacterium was grown in the SCB1 medium (1 g SCB + 10% (v/v) L-tryptophan). This is because the pigment was produced only in the support material and not excreted in the solution. Other medium formulations showed negligible (SCB5) or poor yields of pigment production (in g L^{-1}); SPW1 (0.01), SPW3 (0.07), SPW5 (0.03), BS1 (0.02), BS5 (0.02), BS10 (0.08), M0.1 (0.05), M0.5 (0.03) and M1 (0.03)

Fig. 3.11 Pigment production yield for *C. violaceum* when grown in SCB, SPW, M and BS (supplemented with 10% (v/v) L-tryptophan)

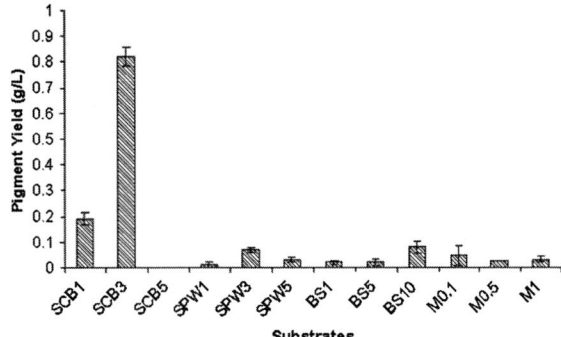

(Fig. 3.11). After 24 h of incubation, the original pale-yellow appearance of SCB and green color of SPW turned to dark purple due to the formation of pigment. Adequate contents of cellulose, hemicellulose and lignin in SCB acted as a good C source as well as a useful adhesion site for bacterial growth and pigment production. However, this process only proceeds in a strictly aerobic condition as demonstrated by the impediment of pigment production in anaerobic condition (SCB5). Similar finding was also reported by DeMoss and Evans (1959).

S. marcescens: Growth of *S. marcescens* in various growth medium was monitored to determine the best medium for bacterial growth and production of pigments. A total of six media were used namely NB (100%), liquid pineapple waste, LPW (100%), brown sugar, BS (1% v/v), BS (10% v/v), BS (1% v/v in LPW) and BS (10% v/v in LPW). Active cultures (20 mL) of *S. marcescens* were inoculated into the respective media (180 mL) and shaken at 200 rpm, 25 °C for 24 h. Each of the samples was complimented with a cell-free control. Figure 3.12 shows the intensity of pigments produced by *S. marcescens* when grown in different media compositions. Based on the intensity of red coloration, highest pigment production was obtained in 10% (v/v) BS (in water), followed by NB, 1% (v/v) BS and 1% (v/v) BS in LPW. No pigment was observed when the bacterium was grown in LPW only or at high volumes of LPW, hence the addition of BS to promote growth and pigment production (Fig. 3.13). Low pigment production in LPW medium was due to the insufficient carbon source present in the medium. The composition of total sugars present in 1 L of LPW was 126.98 ± 0.02 mg which was 1,000 times lower compared to the total sugar present in 1 L of BS i.e., 154.206 ± 10.936 g.

3.4 Effect of Light

S. marcescens: By preventing light penetration (via wrapping of the flask using aluminium foil) during 36 h cultivation of bacteria in NB at 25 °C, a more intense red coloration was obtained compared to culture grown in normal condition i.e.,

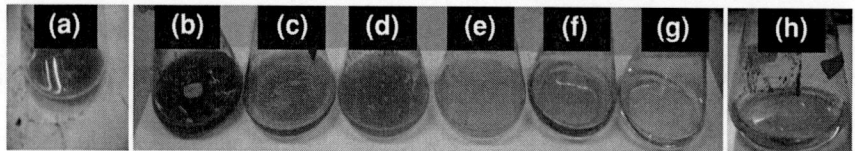

Fig. 3.12 Intensity of pigment produced by *S. marcescens* in different media at 25 °C; (a) NB
(b) 10% (v/v) BS (c) 1% (v/v) BS (d) 1% (v/v) BS in LPW (e) control LPW (f) control 10% (v/v)
BS (g) control 1% (v/v) BS (h) control NB

Fig. 3.13 Production of red
pigment by *S. marcescens* in
different composition of LPW
(in water); (a) 30% v/v
LPW + NB, (b) 50% v/v
LPW + NB (c) 60% v/v
LPW + NB and (d) 100%
v/v LPW

in the presence of light. This can be explained as follows: red pigment is regarded
as a photosensitive pigment, hence it may store energy from light and stimulates its
photo-transformation (Ryazantseva et al. 1995). Photo-transformation enables the
red pigment to undergo association and dissociation reactions into different
compounds i.e., directly affecting its concentration/intensity in the culture media.
Therefore, it can be concluded that the production of red pigment by *S. marcescens*
is affected by light.

3.5 Effect of Cells Immobilization

C. violaceum: SCB was used as support material during the immobilization of
C. violaceum in column system. To allow the SCB surface material to acquire
necessary charge for bacterial attachment, the column was first rinsed with distilled water. An active culture of *C. violaceum* (100 mL) was pumped at a flowrate
of 3.5 mL min^{-1} for 24 h to ensure initial bacterial attachment. Using the same
flowrate, 100 mL of 10% (v/v) L-tryptophan in distilled water (final pH of 7.0) was
pumped continuously for 24 h to promote bacterial growth and pigment production. The pigment produced was recovered using 100 mL of methanol followed by
drying at room temperature. The concentration of pigment was determined
gravimetrically and expressed in g L^{-1}. From the analysis, *C. violaceum* was
capable to produce up to 0.15 g L^{-1} of pigment when immobilized onto support

Fig. 3.14 Intensity of pigment production by *S. marcescens* when grown in LPW supplemented with different concentrations of silica gel. (**a**) LPW without cells (**b**) LPW only, (**c**) 5 g L^{-1}, (**d**) 10 g L^{-1}, (**e**) 15 g L^{-1}, (**f**) 20 g L^{-1} and (**g**) 25 g L^{-1}

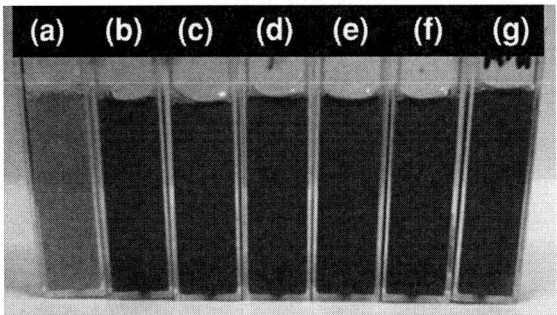

material. Even though this value is substantially lower compared to the concentration of pigment produced in free cells system i.e., 0.82 g L^{-1}, the use of immobilized cells system offers easy recovery of pigment and high cell concentration in a given space. Besides that, the immobilized cells can be maintained under these conditions for a long period without the need for frequent subculturing (Lindsey and Yoeman 1984).

S. marcescens: Cells of *S. marcescens* 10% (v/v) was inoculated into a series of 500 mL Erlenmeyer flasks containing 50 mL of LPW supplemented with silica gel (acted as support material for cells immobilization) followed by incubation (200 rpm, 24 h) at room temperature. The use of silica gel as support material is expected to increase the number of cells per volume ratio, hence increasing the production of prodigiosin (Yamashita et al. 2001). Based on the intensity of red coloration produced in the solution, the addition of 10 g L^{-1} silica gel was judged to contain highest amount of cells, which was responsible for the highest intensity for red coloration (Fig. 3.14).

References

Anuradha VG, Nandini A, Geetha M, Gautam P (2004) A novel medium for the enhanced cell growth and production of prodigiosin from *Serratia marcescens* isolated from soil. BMC Microbiol 4

De Azevedo MBM, Alderete J, Rodriguez JA, Souza AO, Rettori D, Torsoni MA, Faljona–Alario A, Haun M, Duran N (2000) Biological activities of violacein, a new antitumoral indole derivative, in an inclusion complex with ß-cyclodextrin. J Incl Phenom Macrocycl Chem 37:93–101

Durán N, Faljoni–Alario A (1980) Bacterial chemistry-I: studies of a Potential phototherapeutic substance from *Chromobacterium violaceum*. Ann Brazil Acad Sci 52:297–301

Lindsey K, Yeoman MM (1984) The viability and biosynthetic activity of cells of *Capsicum frutescens* Mill. cv. annuum immobilized in reticulate polyurethane. J Exp Bot 35(11): 1684–1696

Lu Y, Wang L, Xue Y, Zhang C, Xing XH, Lou K, Zhang Z, Li Y, Zhang G, Bi J, Su Z (2009) Production of violet pigment by a newly isolated psychrotrophic bacterium from a glacier in Xinjiang, China. Biochem Eng J 43:135–141

DeMoss RD, Evans NR (1959) Physiological aspects of violacein biosynthesis in nonproliferating cells. J Bacteriol 78:583–586

Nancy NG, Gauthier MJ (1979) New prodigiosin-like pigment from alteromonas rubra. Appl
    Environ Biol 37
Ryazantseva IN, Andreyeva IN, Klementyeva GS, Ogorodnikova TI, Petrov VY (1995) Pigment-
    dependent light influence on the energetics of Serratia marcescens. Thermochimica Acta
    251:63–67
Salmiah MN (2006) A manual on textile testing. University Publication Centre (UPENA),
    Universiti Teknologi, MARA
Sivendra R, Lo HS (1975) Identification of Chromobacterium violaceum: pigmented and
    nonpigmented strains. J Gen Microbiol 90:21–23
Vasconcelos ATR, Almeida DF, Hungria M, Guimarães CT, Antônio RV, Almeida FC, Almeida
    LGP, Almeida R, Alves-Gomes JA, Andrade EM, Araripe J, Araujo MFF, Astolfi-Filho S,
    Azevedo V, Baptista AJ, Bataus LAM, Baptista JS, Belo A, van den Berg C, Bogo M, Bonatto
    S, Bordignon J, Brigido MM, Brito CA, Brocchi M, Burity HA, Camargo AA, Cardoso DDP,
    Carneiro NP, Carraro DM, Carvalho CMB, Cascardo JCM, Cavada BS, Chueire LMO,
    Creczynski-Pasa TB, Cunha Junior NC, Fagundes N, Falcão CL, Fantinatti F, Farias IP, Felipe
    MSS, Ferrari LP, Ferro JA, Ferro MIT, Franco GR, Freitas NSA, Furlan LR, Gazzinelli RT,
    Gomes EA, Gonçalves PR, Grangeiro TB, Grattapaglia D, Grisard EC, Hanna ES, Jardim SN,
    Laurino J, Leoi LCT, Lima LFA, Loureiro MF, Lyra MCCP, Madeira HMF, Manfio GP,
    Maranhão AQ, Martins WS, Mauro SMZ, Medeiros SRB, Meissner RV, Moreira MAM,
    Nascimento FF, Nicolas MF, Oliveria JG, Oliveira SC, Paixão RFC, Parente JA, Pedrosa FO,
    Pena SDJ, Pereira JO, Pereira M, Pinto LSRC, Pinto LS, Porto JIR, Potrich DP, Ramalho Neto
    CE, Reis AMM, Rigo LU, Rondinelli E, Santos EBP, Santos FR, Schneider MPC, Seuanez
    HN, Silva AMR, Silva ALC, Silva DW, Silva R, Simões IC, Simon D, Soares CMA, Soares
    RBA, Souza EM, Souza KRL, Souza RC, Steffens MBR, Steindel M, Teixeira SR, Urmenyi
    T, Vettore A, Wassem R, Zaha A, Simpson AJG (2003) The complete genome of
    Chromobacterium violaceum reveals remarkable and exploitable bacterial adaptability. Proc
    Natl Acad Sci USA 100:11660–11665
Yamashita M, Nakagawa Y, Li H, Matsuyama T (2001) Silica gel-dependent production of
    prodigiosin and serrawettins by Serratia marcescens in a liquid culture. Microbes and Environ
    16:250–254

Chapter 4
Application of Bacterial Pigments as Colorant

Abstract In the last decade, investigations about possible use of natural dyes in textile dyeing processes have been carried out by various research groups. Various kind of natural dyes (e.g. *Hibiscus mutabilis*, *Quercus infectoria* and *Cassia tora* L.) were used to dye different types of materials (e.g. cotton, jute, wool, silk and leather) normally in the presence of mordant (e.g. alum, copper sulfate and ferrous sulphate). Studies on the dyeing techniques were attempted using both conventional (alkaline, acidic or neutral baths) and non-conventional methods (ultrasonic, microwave, sonicator and supercritical carbon dioxide fluids). The degree of dyeing was normally compared based on the colorfastness properties which can be defined as the property of a pigment or dye, or materials containing the coloring material, to retain its original hue, without fading, running or changing when wetted, washed, cleaned or stored under normal conditions when exposed to light, heat or other influences. Essentially, this means that different dyes will have different fastness on different materials.

Keywords Pigment · Bacteria · Colorfastness · Prodigiosin · Lightfastness · Violacein · Application

General methods: The bacterial pigments were harvested via centrifugation (8000 rpm, 5 min) from bacterial cultures grown in NB for 24 h at 200 rpm and 30 °C. Among the natural materials and chemicals used as *mordant* during the fabric dyeing process include 70 g L^{-1} alum [$KAl(SO_4)_2 \cdot 12H_2O$], 5 g L^{-1} ferric sulphate [$Fe_2(SO_4)_3$], 5 g L^{-1} copper sulphate ($CuSO_4$), sodium silicate (Na_2SiO_3) and 50 g L^{-1} slaked lime [$Ca(OH)_2$]. Alum was prepared by mixing the solution vigorously for 5 min. The resulting clear solution was used as a mordant while the

W. A. Ahmad et al., *Application of Bacterial Pigments as Colorant*,
SpringerBriefs in Molecular Science, DOI: 10.1007/978-3-642-24520-6_4,
© The Author(s) 2012

Table 4.1 General materials used during the coloring of fabrics using bacterial pigment

Material	Description
Tamarind	Used as mordant; tamarind solution 4% (w/v) was prepared by dissolving 4 g of tamarind in 100 mL of deionized water (Chua 2007)
Soap	The soap solution was prepared by mixing 5 g of standard soap (SDC enterprises limited, UK) with 2 g anhydrous Na_2CO_3 (Hamburg Chemicals) in 1 L distilled water. The mixture was homogenized via heating at 60 °C for 15 min (Salmiah 2006)
Perspiration	The perspiration solution was prepared by dissolving L-histidine monohydrochloride monohydrate (0.5 g), NaCl (5.0 g) and crystallized Na_2HPO_4 (5.0 g) in 1 L of distilled water. The pH was then adjusted to 8.0 using 0.1 N NaOH (Salmiah 2006)
Fabrics	Five types of fabrics were used including natural (cotton and silk) and synthetic fibers (polyester, acrylic and polyester microfiber). The fabrics were obtained from the Program of Textile Technology, Faculty of Applied Sciences, Universiti Teknologi Mara (UiTM), Shah Alam, Malaysia
Candle	Commercial candles were used in candle making. The candles used were fluted (Race Horce, Malaysia) and transparent (Tea Light Candle, China) candles

precipitates were retained for subsequent use. For synthetic dye, 0.35 g of reactive remazol blue or 0.15 g of reactive remazol violet was dissolved in 1 L of distilled water.

The *soap solution* was prepared by mixing 5 g of standard soap with 2 g of anhydrous Na_2CO_3 in 1 L of distilled water (Table 4.1). The solution was heated at 60 °C for 15 min to homogenize the mixture (Salmiah 2006). To prepare the *perspiration solution*, 0.5 g of L-histidine monohydrochloride monohydrate, 5.0 g of NaCl and 5.0 g of crystallized Na_2HPO_4 in 1.0 L of distilled water. The pH of the soap and perspiration solutions was adjusted to 5.5 and 8.0, respectively using either 0.1 M NaOH or 0.1 M HCl (Salmiah 2006).

In this work, two fabric-dyeing methods were attempted namely *boiling* and *brushing* technique. In the boiling method, 1 g of fabric was immersed in 20 mL of bacterial culture broth and heated at either 80 °C (1 h) or 130 °C (1.5 h). Upon cooling, the dyed fabrics were washed with cold water followed by the mordanting process. In this work, the post-mordanting technique, i.e., dyeing followed by mordanting was used during the dyeing of fabrics using bacterial pigment. This technique requires 15 min mordanting followed by washing and drying, all carried out at room temperature. However, a 1.5 h post-mordanting process was required when $NaSiO_3$ was used as mordant. The washing step was carried out using tap water to remove excess mordant on the fabric prior to drying. For the brushing technique, fabrics were dyed using brush that should ensure even distribution of the dye on the fabric materials. This is followed by the mordanting procedure using $NaSiO_3$.

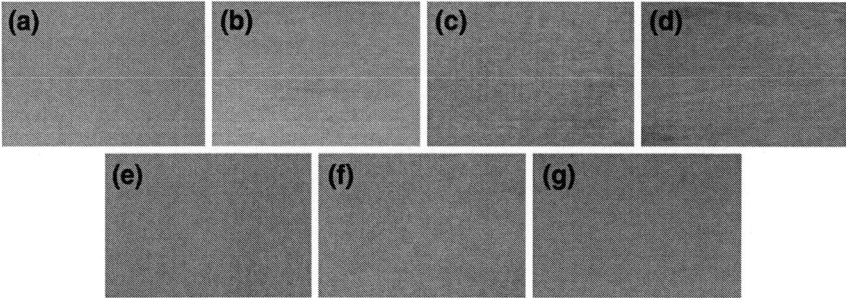

Fig. 4.1 Dyeability of violacein on different fabrics using alum as mordant; (**a**) pure cotton (PC), (**b**) pure silk (PS), (**c**) pure rayon (PR), (**d**) rayon jacquard (RJ), (**e**) silk satin (SS), (**f**) cotton (C) and (**g**) polyester (P)

4.1 Initial Evaluation on Potential Application of Bacterial Pigments as Colorant

4.1.1 Violacein

To evaluate the potential of violacein as coloring material for fabrics, both natural and synthetic fibes were used namely pure cotton (PC), pure silk (PS), pure rayon (PR), jacquard rayon (JR), cotton (C), silk satin (SS) and polyester, P (as the sole synthetic fiber). All fabrics were supplied either by the School of Textile Technology, Faculty of Applied Sciences, Universiti Teknologi Mara (UiTM), Shah Alam or by the Malaysian Craft Development Corporation, Kota Bharu, Kelantan. The fabrics were dyed using the *boiling* technique with alum used as mordant. Results obtained showed the capability of violacein to color both types of fibers with varying degrees of shades and intensity (Fig. 4.1).

Intense colorations on PR, RJ and SS indicate its suitability to be dyed with violacein compared to C, PC, PS and P. Different color shades obtained during dyeing were due to different rates of adsorption between dye–dye and fiber–fiber. In the fabric dyeing process, the chemical nature of different textile materials is important as this will determine the exact mechanism by which dye is adsorbed onto particular reactive groups on the fabrics. This can be explained as follows; upon entering the fiber, dye molecules would be gradually transported from the aqueous phase onto the fiber (Vickerstaff 1954). Rayon (pure and jacquard) and cotton (pure and normal) exhibited different responses toward dyeing even though both are made up of cellulosic materials. This is because during the manufacturing of viscose rayon, dissolution of the natural cellulose corresponds to a very intensive swelling and leads to a much greater micellar surface in the finished rayon compared to that in the original cellulose. At the same time oxidative degradation occurs, leading to an increased number of carboxyl groups in the rayon. Since this carboxyl group will form negative charge in water, the pigments

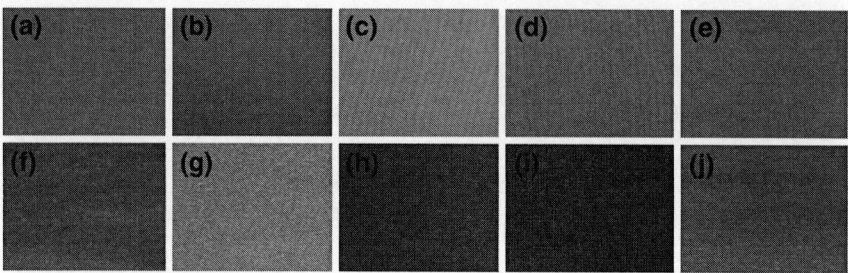

Fig. 4.2 Dyeability of violacein on pretreated cotton and silk fibers; (**a**)cotton without mordant, (**b**) cotton + alum, (**c**) cotton + $Fe_2(SO_4)_3$, (**d**) cotton + $CuSO_4$, (**e**) cotton + $Ca(OH)_2$, (**f**) silk satin without mordant, (**g**) silk satin + alum, (**h**) silk satin + $Fe_2(SO_4)_3$, (**i**) silk satin + $CuSO_4$ and (**j**) silk satin + $Ca(OH)_2$

become attached by the formation of hydrogen bond to the COO^- group in the fiber with OH or NHR group present in the pigment, resulting in good coloration of the rayon-based textile material (Vickerstaff 1954). It is well known that a hydrogen atom can act as an electron acceptor particularly when it is directly attached to either nitrogen or oxygen atoms. With higher presence of electron-donating and electron-accepting groups between the fiber and pigment, more pigment will be attached to the fiber, hence increasing the dyeing ability (Vickerstaff 1954).

The pretreatment of fabric was also carried out before the dyeing process in order to remove natural impurities present on the cotton and silk fabrics. Some examples of impurities that may be present include fats, waxes, pectins and related substances, minerals, heavy metals, amino acids or proteins, lubricants and knitting oils. In the present work, both the pretreated cotton (via the scouring method using a solution containing a mixture of Na_2CO_3 and alum) and silk satin (degumming process using alum solution) showed higher color intensity compared to the untreated fabrics (Fig. 4.2). Pretreatment increases swelling of fiber which in turn releases impurities, hence resulting in the liberation of more binding space for the pigment (violacein) on the fiber (Karmakar 1999).

4.1.2 Yellow Pigment

The yellow pigment was extracted via the boiling method from *Chryseobacterium* sp., a locally isolated Gram-negative bacterium. Several types of fabrics namely Natural silk, Dubai silk, Linen, Japanese cotton and Indian cotton were dyed using the yellow pigment by immersing the fabrics in the pigment solution at 80–90 °C. Optimum Color intensity was obtained after 60 min (Fig. 4.3).

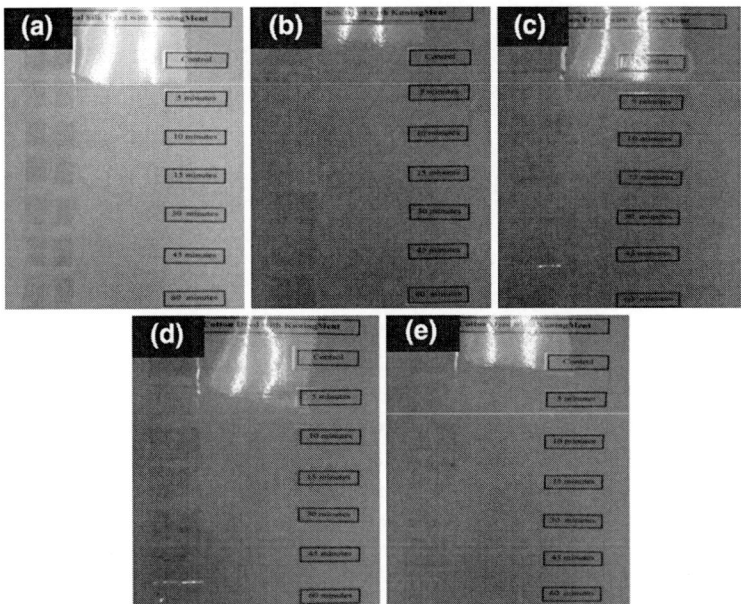

Fig. 4.3 The effect of different immersion time on Color intensity on fabrics for (**a**) natural silk, (**b**) Dubai silk, (**c**) Linen, (**d**) Japanese cotton and (**e**) Indian cotton; no mordant was used

Table 4.2 Dyeing of different fabrics using prodigiosin

	Acrylic	Polyester microfiber	Polyester	Silk	Cotton
Before dyeing					
After dyeing					

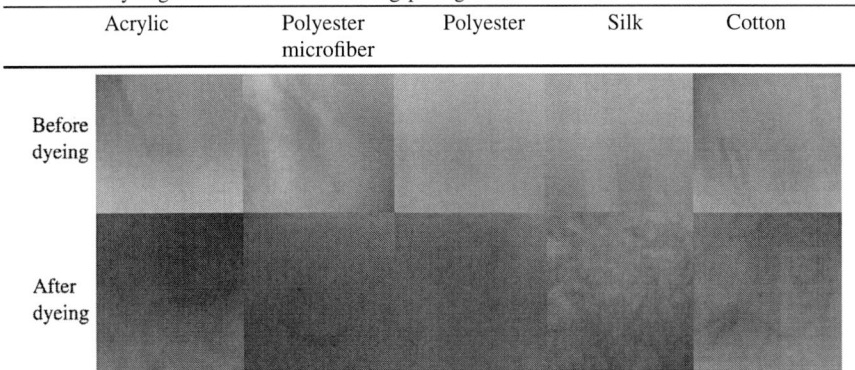

4.1.3 Prodigiosin

The pigment was used to color five types of fabric namely acrylic, polyester microfiber, polyester, silk and cotton. The dyeing method used in this experiment was direct boiling of the fabrics with the bacterial cells (Table 4.2).

Generally, prodigiosin was able to color both the natural and synthetic fibers. However, different color intensity was obtained due to the nature of each fiber which requires certain types of dye materials to produce intense coloration (Vickerstaff 1954). In this work, prodigiosin gave the most intense coloration with acrylic (bright red) followed by Polyester microfiber (dark red), polyester (darker red) and silk (light red uneven). However, the color intensity on cotton fiber was very poor. In dyeing, the chemical natures of different textile materials are important as it affects the exact mechanism by which the dyes are absorbed. Since cotton is made from cellulosic materials, it requires dye molecule which contain groups capable of forming hydrogen bond with the hydroxyl group on the cellulose molecule. The dyeing of cotton presented considerable difficulty as the number of dyes having any attraction for cotton was small (Vickerstaff 1954) and as a result, almost no dyeing occurred on the cotton fabric. Acrylic fabric is made from polyamide fiber. From the FT–IR result, it shows that the pigments consisted of amine functional group at 3292 cm^{-1}, thus in acrylic dyeing process, the pigments become attached by the formation of hydrogen bond to the amide group in the fiber, resulting in good coloration of the acrylic fabric (Vickerstaff 1954).

4.2 Colorfastness Properties of Bacterial Pigments

General methods carried out for the colorfastness tests

Colorfastness can be defined as the property of a pigment or dye, or materials containing the coloring material such as leather, cloth, paper and ink, to retain its original hue, without fading, running or changing when wetted, washed, cleaned or stored under normal conditions when exposed to light, heat or other influences. Essentially, this means that different dyes will have different fastnesses on different materials. For example, linen is much harder to be dyed as opposed to silk or cotton, although indigo dyes both cotton and linen well. A dye which works well on leather will probably not be suitable for wool. Running occurs principally during washing and exposure to detergents and solvents. A dye will run if it has a weak affinity for the material it is attached to, or a much stronger affinity for a non-aqueous solvent. Detergents may cause running because they help to stabilize the hydrophobic regions of dye molecules due to their ability to form micelles. Fading is caused by the chemical alteration of unstable dye molecules to a less strongly colored or colorless form. This is often caused by the action of sunlight, or by the oxidizing action of the atmosphere. The UV radiation in sunlight has enough energy to cause unstable bonds to break or reform. Oxygen and atmospheric water will react with unstable bonds to alter the structure and affect its color. The basic principle for colorfastness testing is to compare the tested fabrics with the Grey Scale for change in color and staining. The Standard Methods and equipments used during the colorfastness evaluation are as listed in Table 4.3.

Colorfastness to washing: A 10 cm × 4 cm of dyed fabric was prepared and weighed. The dyed fabrics were attached with control, i.e., un-dyed fabrics, and

Table 4.3 List of Standard Method, colorfastness test and equipment involved to evaluate colorfastness of fabrics dyed using bacterial pigment

Colorfastness test	Standard Method	Equipment
Washing	MS ISO 105-A05-2003	Color matching cabinet
	MS ISO 105-A04-2003	Staining
Perspiration	MS ISO 105-E04-1996	Perspirometer
	MS ISO 105-A05-2003	Color matching cabinet
	MS ISO 105-A04-2003	Staining
Light	MS ISO 105-B02-2001	Light fastness tester
Rubbing/Crocking	MS ISO 105-X12-2001	Crockmeter
	MS ISO 105-A04-2003	Staining

placed inside a jar filled with standard soap solution with a ratio of 1 g fabric: 50 mL of soap solution. Then, the jar was placed inside the Auto-Wash equipment (Labtec) and allowed to wash for 30 min at 60 °C. After that, the sample was washed and dried under mild sunlight. The effect of washing on the dyeing ability of pigment was determined by comparing the color change of the dyed fabric with white fabric according to the Grey Scale Standard of 1 to 5 (American Association of Textile Chemists and Colorists).

Colorfastness to light: A 114 mm × 50 mm of cardboard was prepared. The dyed fabrics were cut into 50 mm × 10 mm and placed (stapled) horizontally, one above the other on the same cardboard. Blue wool standards (Nos.1–8) were cut into strips of 50 mm × 10 mm and were placed one above the other on a separate cardboard. Both specimens were placed in the sample holder. The sample holder allowed both ends of the standards or samples to be exposed to light. The central area was covered by the framework; thus, the area was not exposed to light. Both samples and standards were exposed to light in Light Fastness Tester (Model No. 225, Halifax, England) for 24 h (Salmiah 2006).

Colorfastness to rubbing/crocking: Four rubbing cotton cloths (50 mm × 50 mm) were prepared for each fabric. Each cloth was fixed to the rubbing finger. After that, two warp and two weft of dyed fabrics for each sample was prepared (200 mm × 110 mm), one warp and weft sample for dry rubbing, and one warp and one weft for wet rubbing. For dry rubbing, the cotton cloths were rubbed using a crockmeter on the dry-dyed fabric sample ten times forward and backward in a straight line along a track of 100 mm in length. The same procedure was applied for wet rubbing except, the cotton cloth was wetted using distilled water before the rubbing on the dyed fabric sample. Lastly, the result was evaluated by comparing the staining on the white rubbing cotton cloth with the Grey Scale Standards of 1 to 5 (Salmiah 2006).

Colorfastness to perspiration: A 60 mm × 60 mm for each dyed fabric was prepared and weighed. The perspiration solution was poured into a beaker containing the dyed fabric. The liquor ratio is 1 g of fabric to 50 mL of perspiration solution. Then, the sample was stirred occasionally for 30 min before being removed from the solution and wringed in the perspirometer. The perspirometer was left in the oven at 37 °C for 4 h. After 4 h the results were evaluated by

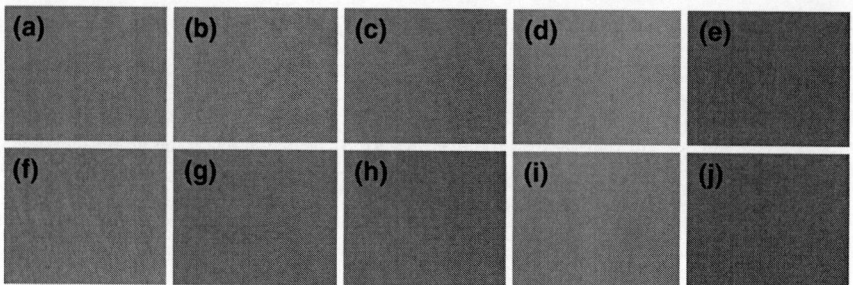

Fig. 4.4 Effect of mordant on the dyeing of cotton and silk satin using violacein; (a) cotton only (b) cotton + alum, (c) cotton + Fe$_2$(SO$_4$)$_3$, (d) cotton + CuSO$_4$, (e) cotton + Ca(OH)$_2$, (f) silk satin only, (g) silk satin + alum, (h) silk satin + Fe$_2$(SO$_4$)$_3$, (i) silk satin + CuSO$_4$ and (j) silk satin + Ca(OH)$_2$

comparing the change in Color of the fabrics and staining of the undyed fabric with the respective Grey Scale Standards of 1 to 5 (Salmiah 2006).

Colorfastness to water: One sample (60 mm × 60 mm) for each dyed fabric was prepared and weighed. Distilled water was dropped on the dyed fabric. The dyed fabric was left in the oven at 37 °C for 4 h. After 4 h the results were evaluated by comparing the change in color of the fabrics and staining of the undyed fabric with the respective Grey Scale Standards of 1 to 5 (Salmiah 2006).

4.2.1 Violacein

For silk satin and cotton, four parameters were varied namely types of mordant (alum, FeSO$_4$, CuSO$_4$, NaSiO$_3$ and Ca(OH)$_2$), concentration of mordant (5 g L^{-1} and 10 g L^{-1}), pretreatment of fabrics and fastness properties. The fabric pretreatment step was carried out to remove pectic substances and cotton wax contained in cotton woven fabrics as well as substances present on the surface of silk fabrics. Cotton fabric (1 g) was pretreated using either 15 g L^{-1} alum or 0.18 g L^{-1} Na$_2$CO$_3$ via 2 h of boiling. Then, the mixture was left to cool for 24 h followed by washing and drying at room temperature. Similar procedure was applied for silk, however without the addition of Na$_2$CO$_3$. Natural dyes often require metallic mordant to increase the affinity between fiber and dye as well as preventing the color to fade from washing steps or exposure to light (Chairat et al. 2007). Figure 4.4 shows the color of cotton and silk satin fabrics dyed with four different metals as mordant, i.e., Al (alum), Fe, Cu and Ca.

The use of slake lime mordant (Ca(OH)$_2$) resulted in darker color compared to the control fabrics for both cotton and silk satin. Gold-colored fiber was obtained when Fe$_2$(SO$_4$)$_3$ was used while CuSO$_4$ resulted in grey coloration. However, the effect of alum was insignificant. The fastness properties of cotton and silk dyed with violacein satin are shown in Table 4.4. Overall, the fastness properties of both

Table 4.4 Shades and fastness properties of violacein - dyed cotton and silk satin in the presence of various mordants

Fabric	Mordant	Colorfastness										
		Light	Washing		Rubbing/crocking				Perspiration				Water
			AC	S	Dry		Wet		A		B		CC	AC
					wf	wp	wf	wp	CC	AC	CC	AC	
Cotton	No	1	2/3	3/4	4	4/5	4/5	4/5	5	5	5	5	5	5
	Alum	1	4	4	4	4	4/5	4/5	5	5	5	5	5	5
	Fe$_2$(SO$_4$)$_3$	2	2	4/5	4	4	4	4	3/4	4/5	3	4/5	4/5	4/5
	CuSO$_4$	2/3	2/3	4	4/5	4/5	4/5	4	3	4/5	3	4/5	4/5	4/5
	Ca(OH)$_2$	1	3	3	4	4	4/5	4/5	5	4/5	5	4/5	5	4/5
Silk Satin	No	1	3/4	3	4/5	4/5	4/5	4/5	5	5	5	5	5	5
	Alum	2	4	4	4/5	4/5	4/5	4/5	3/4	4/5	3	4/5	4/5	4/5
	Fe$_2$(SO$_4$)$_3$	1	2	4/5	4	4	4/5	4	3	4/5	3/4	4/5	4/5	4/5
	CuSO$_4$	1	3	4	4/5	4/5	4/5	4/5	3/4	4/5	4	4/5	4/5	4/5
	Ca(OH)$_2$	2	3/4	3/4	4/5	4/5	4/5	4/5	5	5	5	5	5	5

AC assessing Color, *S* staining on white fabric, *wf* weft, *wp* warp, *A* acidic condition, *B* basic condition, *CC* Color change; *1* very poor, *2* poor, *3* fair, *4* good, *5* excellence

Table 4.5 Cotton and silk satin after dyed using violacein in the presence of 5 and 10 gL^{-1} Fe$_2$(SO$_4$)$_3$ and CuSO$_4$ (as mordant)

	Fe$_2$(SO$_4$)$_3$		CuSO$_4$	
	5 g L^{-1}	10 g L^{-1}	5 g L^{-1}	10 g L^{-1}
Cotton				
Silk satin				

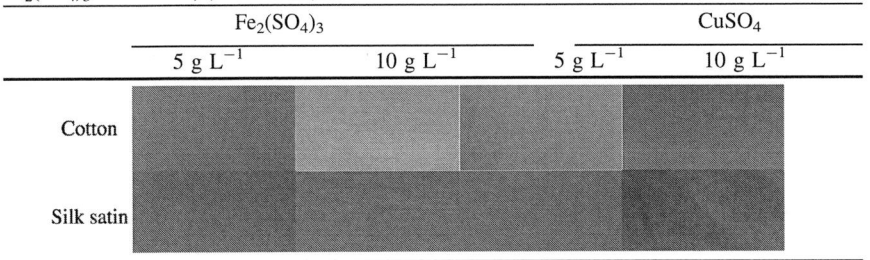

fabrics were rated from 3 (fair) to 5 (excellence) except for lightfastness. The intensity of violacein on cotton and silk satin can be enhanced via increasing the concentration of mordant (Fe$_2$(SO$_4$)$_3$ and CuSO$_4$), as depicted in Table 4.5. Increased coloration intensity at higher mordant concentration can be attributed to the increased complex formation between fiber and dye molecules resulting from the high concentration of metal ions present in the solution (from the mordant). However, the fastness properties between dyed fabrics applied with and without mordant did not show any significant difference (Table 4.6). This may be due to reactive sites present which is responsible to bind metal mordants. When the site is occupied by a metal mordant, it is no longer available to attract more metal mordant, so that only a monomolecular layer on the fiber surface can be obtained.

Table 4.6 Shade and fastness properties of dyed cotton and silk satin mordanted using different concentration of mordants

Fabric	Mordant	Concentration	Colorfastness											
			Light	Washing		Rubbing/crocking				Perspiration				Water	
				AC	S	Dry		Wet		A		B		CC	AC
						wf	wp	wf	wp	CC	AC	CC	AC		
C	Fe₂(SO₄)₃	5 g L⁻¹	2	2	4/5	4	4	4	4	3/4	4/5	3	4/5	4/5	4/5
SS			1	2	4/5	4	4	4/5	4	3	4/5	3/4	4/5	4/5	4/5
C	CuSO₄		2/3	2/3	4	4/5	4/5	4/5	4	3	4/5	3	4/5	4/5	4/5
SS			1	3	4	4/5	4/5	4/5	4/5	3/4	4/5	4	4/5	4/5	4/5
C	Fe₂(SO₄)₃	10 g L⁻¹	1	3	4	4	4	4	4/5	4	4/5	3/4	4/5	4	4
SS			2/3	3	4	4	4	4/5	4	3/4	4/5	3	4/5	4/5	4/5
C	CuSO₄		1	3/4	4/5	4/5	4/5	4	4	4	4/5	4/5	4/5	4/5	4/5
SS			2/3	3/4	4	4/5	4/5	4	4	3	4/5	3/4	4/5	4/5	4/5

C cotton, *SS* silk satin, *AC* assessing Color, *S* staining on white fabric, *wf* weft, *wp* warp, *A* acidic condition, *B* basic condition, *CC* Color change; *1* very poor, *2* poor, *3* fair, *4* good, *5* excellence

The excess metal mordant may leach-out during the fastness test since it does not have any interaction with the fabrics (Vickerstaff 1954).

The stark difference between fabrics which were dyed using violacein, with and without the presence of mordant, can be attributed to the presence of metallic compound from the mordant, i.e., Fe and Cu. Being a transition element, Fe and Cu have the ability to change color, from its original state of green (Fe^{3+}) and blue (Cu^{2+}) to brown/grey (Fe) and dark blue (Cu), when it complexes with suitable ligand such as water (Rodger 1994). The coordination between transition metals and colored organic ligand resulted in a compound with better lightfastness than that of the ligand only. This was due to the reduction in the electron density of the chromophore, which in turn leads to improved resistance to photochemical oxidation (Christie 2001). In this work, good colorfastness of fabric to washing was obtained when alum was used as the mordant. The presence of Al^{3+} ions (role as electron acceptor) would lead to the formation of coordination binding with the violacein molecule, followed by the formation of an insoluble complex, hence increasing the binding affinity of violacein to the fibers. From these findings, it can be concluded that violacein can act both as direct dye and mordant dye based on its ability to adhere to the fabric molecules with and without the presence of mordant. However, better colorfastness was obtained in the presence of mordant. One important point to note is the slightly lower color intensity for the fabric when dyed using violacein compared to reactive dyes. Even though, both the cotton and silk satin were dyed using violacein in the presence of mordants (alum and $Ca(OH)_2$), color produced was slightly less intense compared to reactive dye used, i.e., Na_2SiO_3 (Fig. 4.5).

This clearly indicates the higher affinity of reactive dye toward cotton and silk satin fabrics compared to violacein as well as strengthening the notion that natural pigment would produce a mild color shade (Samanta and Agarwal 2009). This

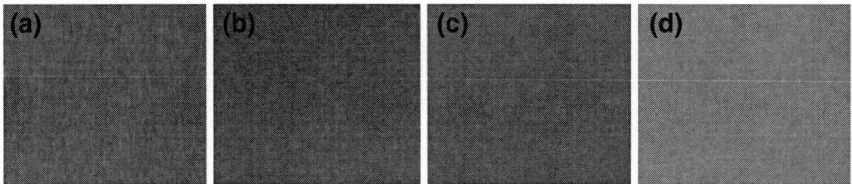

Fig. 4.5 Color intensity for (**a**) cotton + Na₂SiO₃ (reactive dye), (**b**) silk satin + Na₂SiO₃ (reactive dye), (**c**) cotton + violacein + alum and (**d**) cotton + violacein + Ca(OH)₂

Table 4.7 Comparison of fastness properties of natural dyes with reactive dyes

Fabric	Dye	Mordant	Colorfastness												
			Light	Washing	Rubbing/crocking				Perspiration				Water	
				AC	S	Dry		Wet		Acid		Alkaline		CC	AC
						wf	w	wf	w	CC	AC	CC	AC		
Cotton	Vio	Alum	1	4	4	4	4	4/5	4/5	5	5	5	5	5	5
	R.dye	Na₂SiO₃	4	4	4	4	4	4	4/5	5	4/5	5	4/5	5	4/5
Silk Satin	Vio	Ca(OH)₂	2	3/4	3/4	4/5	4/5	4/5	4/5	5	5	5	5	5	5
	R.dye	Na₂SiO₃	3	4	4	4/5	4/5	4	4	5	4/5	5	4/5	5	4/5

Vio violacein, *R. dye* reactive dye, *AC* assessing color, *S* staining on white fabric, *wf* weft, *w* warp, *CC* color change; *1* very poor, *2* poor, *3* fair, *4* good, *5* excellence

condition can also be supported from the colorfastness profile, as shown in Table 4.7. Highest colorfastness properties was obtained for cotton dyed with reactive dye, with average rating of 4/5 followed by cotton dyed with violacein and alum (4/5), silk satin + reactive dye (4) and silk satin + violacein and Ca(OH)₂. However, one significant advantage of reactive dye was its higher lightfastness ratings of 4 (cotton) and 3 (silk satin) compared to violacein with ratings of 1 (cotton) and 2 (silk satin). The higher light fastness properties for reactive dyes can be attributed to the strong intramolecular H-bonding which exists in the form of six membered rings. This enhances the stability of the compound by reduced electron density at the chromophore. As a result, sensitivity of dye toward photochemical oxidation was reduced (Ali et al. 2009).

4.2.2 Prodigiosin

The colorfastness test was carried out using standard soap solution and evaluated for color staining and color change by comparing color intensity of five fabrics ,i.e., polyester, acrylic, cotton, polyester microfiber and silk, before and after washing procedure using the Grey Scale for Assessing Change in Color (MS ISO 105-A05-2003) and Grey Scale for Assessing Staining (MS ISO 105-A04-2003).

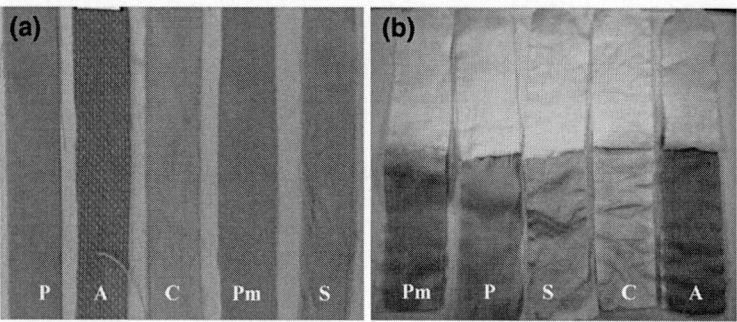

Fig. 4.6 The Color of fabric dyed with prodigiosin before (**a**) and after (**b**) washing; *P* polyester, *A* acrylic, *C* cotton, *Pm* polyester microfiber, *S* silk

Table 4.8 Colorfastness properties (color and staining) for fabrics dyed with prodigiosin

Change in Color				Change in staining			
Fabrics	1st Person	2nd Person	Average	Fabrics	1st Person	2nd Person	Average
S	2	2/3	2	S	5	5	5
P	2/3	2/3	2/3	P	5	5	5
C	1	1	1	C	5	5	5
A	5	5	5	A	5	5	5
Pm	3	4	3/4	Pm	5	5	5

Based on the evaluation (Fig. 4.6 and Table 4.8), acrylic fiber dyed with prodigiosin showed excellent colorfastness toward washing with a rating of 5, followed by polyester microfiber (3/4), polyester (2/3), silk (2) and cotton (1). This is because acrylic fiber provided better force of interaction with the dye and reduced its tendency to be washed out on laundering. Overall, high colorfastness (staining) of 5 indicates that prodigiosin dyed on the fabrics would not stain other fabrics.

The lightfastness (colorfastness to light) properties were determined using the Light Fastness Tester. This method was used to simulate the impact of natural outdoor sunlight on colored fabrics. Dyed fabrics and the blue wool standards were simultaneously exposed to light for 24 h. The blue wool standards ranged from No. 1 (very low lightfastness) to No. 8 (very high lightfastness) each standard being approximately twice as fast as the one below its position (Salmiah 2006). Prodigiosin exhibited good lightfastness properties on acrylic (Fig. 4.7) with a rating of 3 (based on AATCC grey scale for color change) while the rest of the fabrics showed poor lightfastness properties with a rating of 1. These findings further strengthen earlier notion of the instability of bacterial pigment or any other natural pigments toward light compared to synthetic pigment.

Results from the evaluation of colorfastness against perspiration (Fig. 4.8) showed that acrylic, silk and polyester displayed good colorfastness with ratings between 4 and 5 while cotton and polyester microfiber had average ratings of 3 and

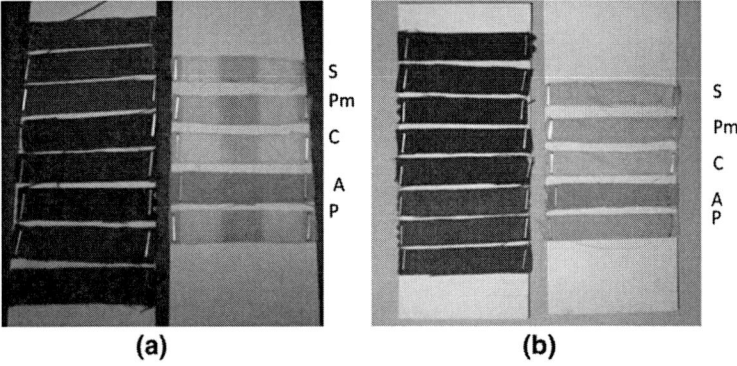

Fig. 4.7 Lightfastness properties of the fabrics (**a**) before and (**b**) after exposure to light (Light Fastness Tester) for 24 h; *S* silk, *Pm* polyester microfiber, *C* cotton, *A* acrylic and *P* polyester

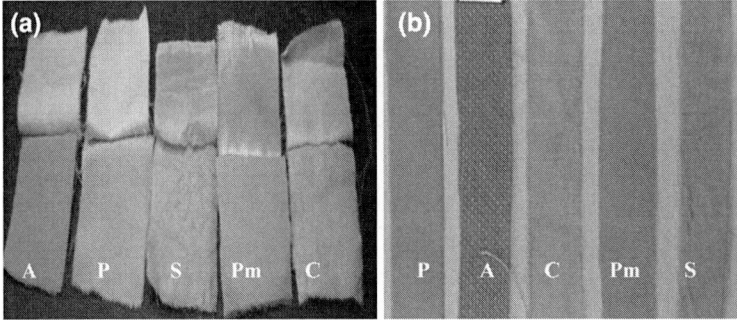

Fig. 4.8 Colorfastness to perspiration properties for fabrics (**a**) before and (**b**) after determination using a perspirometer; *P* polyester, *A* acrylic, *C* cotton, *Pm* polyester microfiber, *S* silk

Table 4.9 Rubbing/crocking colorfastness properties (staining) for fabrics dyed with prodigiosin

Types of fabric	Wet		Dry	
	Warp	Weft	Warp	Weft
Silk	5	5	5	5
Acrylic	4/5	4	4/5	5
Polyester	4/5	5	5	5
Polyester microfiber	4/5	4/5	4	5
Cotton	4/5	4/5	5	5

3/4. All fabrics showed good rating of 5 for staining. The colorfastness to rubbing/crocking was tested using a crockmeter, both in wet and dry conditions. High ratings of 4 and 5 were obtained for the change in staining (Grey scale). This indicated that minimum amount of prodigison was transferred from the dyed fabrics onto the test fabric (Table 4.9).

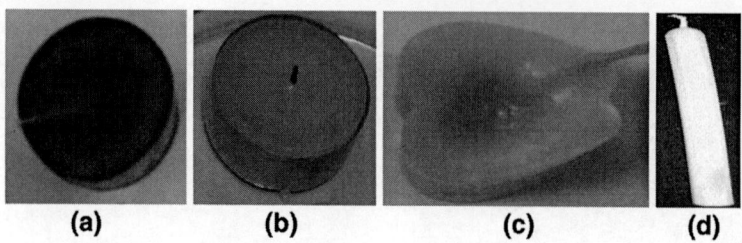

| (a) | (b) | (c) | (d) |

Fig. 4.9 Application of prodigiosin in (**a**) translucent candle and (**c**) fluted candle; (**b**) original translucent candle (**d**) original control fluted candle

4.3 Application in Candle and Paper Making

Commercial candles (fluted and translucent) were placed in a beaker and heated until completely melted before the addition of the bacterial culture broth. The mixtures were homogenized and poured into the mold. The wicks were immediately placed into the centre of the mold and the candles were left to cool at room temperature for 1 h. Translucent candle showed a more intense coloration compared to the fluted candle (Fig. 4.9).

To evaluate the potential application of prodigiosin in paper making, bacterial culture broth (3 mL) and one teaspoon of cornstarch were homogeneously blended in one half of the thick pulp, whereas the other half was not added with pigment (control). The pulp was then evenly spread onto a net to drain the excess water followed by 24 h drying (Martinko and Mardigan 2006). The prodigiosin - dyed paper was exposed to sunlight and fluorescent light for 4 h while paper unexposed to light acted as control. Initial intense red coloration on the paper was substantially reduced to light red upon exposure to both sunlight and fluorescent light, with sunlight exerting the higher fading effect (Fig. 4.10). This effect can be attributed to wider range of light wavelengths for sunlight compared to the fluorescent light (Scheyer and Chiweshe 2001).

4.4 Application in Batik-Making

"Batik", which is also known as the textile art of Malaysia, is a popular gown-like dress worn mostly by woman in the South East Asian region. The most popular motifs include leaves, flowers and geometrical designs.

Pigments produced by *S. marcescens*, *C. violaceum* and *Chryseobacterium* sp. used either 95% (v/v) methanol (*S. marcescens* and *C. violaceum*) or 99.5% (v/v) acetone (*Chryseobacterium* sp.) at a ratio of 1:5 (supernatant) or until the pellet was colorless. This was followed by rotary evaporation to concentrate the pigment. The batik dyeing was carried out on creeps, satin and silk fabrics. The chosen

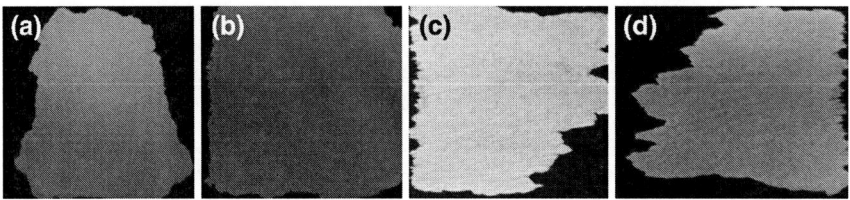

Fig. 4.10 Application of prodigiosin on paper; (**a**) original paper (control), (**b**) paper dyed with prodigiosin, (**c**) dyed paper exposed to sunlight and (**b**) dyed paper exposed to fluorescent lamp

pattern was first drafted onto the fabric by a "Batik-Tulis" maker i.e., the painter, using pencil. Then, melted wax (mixture of beeswax and paraffin wax) was applied over the drafted motifs using a technique called "*canting*", pronounced as "chanting". The beeswax holds the fabric while paraffin wax will allow cracking, which is a typical characteristic of batik. Wherever the wax has seeped through the fabric, the dye will not penetrate. After waxing process, the fabrics were dyed using the extracted bacterial pigments using the brushing technique. The colour tone was adjusted by adding either ethyl acetate (for red and purple pigment) or acetone (yellow pigment). This was followed by immersing the fabrics into boiling water containing fixer (alum, $Fe_2(SO_4)_3$, $CuSO_4$ or $Ca(OH)_2$) to remove excessive wax as well as fixing the bacterial pigments onto the fabrics. The "batik" was then let to dry under mild sunlight (Fig. 4.11).

4.5 Application as Ink

The bacterial pigments were evaluated for its potential role as ink in ballpoint pen and highlighter pen. There are two types of ballpoint-pen ink namely oil-based ballpoint-pen ink and water-based ballpoint-pen ink. The basic components in ballpoint-pen inks are coloring agent, solvent and resin. Dyes and pigments which are soluble or dispersible in aqueous media can be used as coloring agents in inks. Besides the basic components, several other compounds were also added to the inks as additives that include amine derivatives as the pH-controlling agent or mildew-proofing agent, fluorine-containing surfactant that is responsible to increase solvent penetrability as well as defoaming agent, rustproofing agent and lubricant. The addition of shear-viscosity reducing agent such as cross-linked acrylic resin and fatty acid metal salt can prevent the leakage of ink due to the gap between the ball and the tip when the pen is not used. Typical ink for the highlighter pen would consist of liquid vehicle, colorant and acidic buffer solution. Liquid vehicle is the major component in the ink and is used to carry the other highlighter ink component to the substrate. Liquid substrate can be of any liquid type including surfactant, solvent, co-solvent, buffer, biocide, viscosity modifier, stabilizing agent, complexing agent and water. To evaluate the role of bacterial

Fig. 4.11 Application of bacterial pigments on fabrics during the "batik"-making process; (**a**) drafting of motifs using pencil, (**b**) applying wax over the pencil-drawn motifs using the "canting" technique, (**c**) colouring process using bacterial pigments, (**d**) dewaxing and fixation process, (**e**) drying of "batik" under mild sunlight and (**f**) finished product

pigments as ink in ballpoint pen, three types of test were carried out namely the *Ink-Rubbing* test, *Ink-Follow* test and *Stability of Ink Toward Light* test. Similar tests were carried out for the highlighter pen except the *Ink-Follow* test was replaced by the *Ink-Drop* test.

During the ballpoint-pen test, different ink compositions were prepared as described in Tables 4.10 and 4.11.

From the results, it was clearly shown that color intensity increases with amount of pigment added. The violacein Ink 4 gave dark purple (Fig. 4.11d) while the

Table 4.10 Different ink composition of violacein during the ballpoint- pen test

Component	Violacein ink 1	Violacein ink 2	Violacein ink 3	Violacein ink 4
Dry violacein (wt%)	0.00	0.25	1.24	3.05
Gum A\vrabic (wt%)	18.91	3.73	18.67	18.33
Ethanol (wt%)	27.53	40.75	27.19	26.69
Distlilled water (wt%)	44.12	53.41	43.56	42.76
Iota-carageenan (wt%)	4.45	1.86	9.33	9.16
Total (wt%)	100	100	100	100

Table 4.11 Different ink composition of prodigiosin during the ballpoint-pen test

Component	Prodigiosin ink 1	Prodigiosin ink 2	Prodigiosin ink 3	Prodigiosin ink 4
Dry prodigiosin (wt%)	0.00	0.56	0.55	3.05
Gum arabic (wt%)	18.91	18.80	19.89	18.33
Ethanol (wt%)	27.53	27.38	26.72	26.69
Distlilled water (wt%)	44.12	43.86	42.81	42.76
Iota-carageenan (wt%)	4.45	9.40	10.03	9.16
Total (wt%)	100	100	100	100

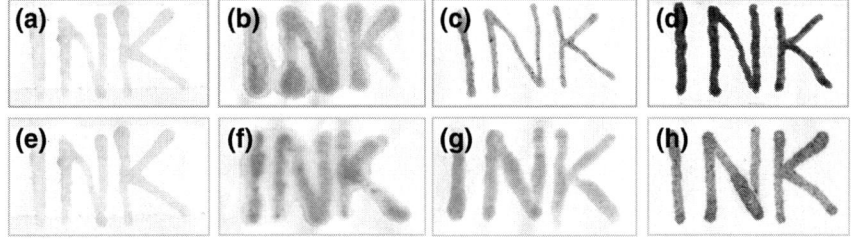

Fig. 4.12 Writing of the word '*ink*' using different compositions of violacein or prodigiosin; (**a**) violacein ink 1, 0% wt., (**b**) violacein ink 2, 0.25% wt., (**c**) violacein ink 3, 1.24% wt., (**d**) violacein ink 4, 3.05% wt., (**e**) prodigiosin ink 1, 0% wt., (**f**) prodigiosin ink 2, 0.25% wt., (**g**) prodigiosin ink 3, 1.24% wt and (**h**) prodigiosin ink 4, 3.05% wt

prodigiosin Ink 4 (Fig. 4.12h) resulted in intense red coloration. The use of excess solvents would result in smearing of the ink (Fig. 4.11b and Fig. 4.11f) which may not be suitable for ballpoint application.

The potential of violacein as ink for highlighter pen was evaluated by preparing different compositions of the ink in the presence of citric acid and glycerol (Table 4.12). The ink was then used to highlight printed words as shown in Fig. 4.13. The effect of solvent used i.e., ethyl acetate and glycerol, on violacein as highlighter ink was also investigated (Fig. 4.14). Based on the smearing effect,

Table 4.12 Different ink composition of violacein during the highlighter test

Ink Component	Ink A	Ink B	Ink C	Ink D	Ink E
Dry violacein (wt%)	1.98	1.39	0.87	0.67	0.60
Citric acid (wt%)	1.33	0.93	0.59	0.45	0.41
Glycerol (wt%)	96.69	97.68	98.54	98.88	98.99
Total (wt%)	100	100	100	100	100

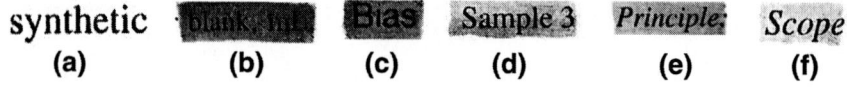

Fig. 4.13 Highlighted words using different compositions of violacein as ink; (**a**) without highlighter, (**b**) Ink A, 1.98% wt., (**c**) Ink B, 1.39% wt., (**d**) Ink C, 0.87% wt., (**e**) Ink D, 0.67% wt. and (**f**) Ink E, 0.60% wt

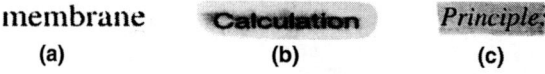

Fig. 4.14 Effect of solvent on violacein as ink in highlighter pen; (**a**) without highlighter, (**b**) ethyl acetate and (**c**) glycerol

glycerol performed better as solvent compared to ethyl acetate. Although solvent can include any liquid capable to transfer the colorant and acid buffer to the substrate, a good solvent is a liquid which can evaporate in short time and having low degree of smear.

References

Ali S, Hussain T, Nawaz R (2009) Optimization of alkaline extraction of natural dye from henna leaves and its dyeing on cotton by exhaust method. JCLP 17:61–66

Chairat M, Bremner JB, Chantrapromma K (2007) Dyeing of cotton and silk yarn with the extracted dye from the fruit hulls of mangosteen. Garcinia mangostana linn Fibres and Polymers 8(6):613–619

Christie RM (2001) Colour chemistry. Royal Society of Chemistry, United Kingdom, pp 12–44

Karmakar SR (1999) Chemical technology in the pre-treatment processes of textiles. Elsevier Science, The Netherlands

Martinko JM, Madigan MT (2006) Brock: Biology of microorganism, 11th edn. Pearson Education International, USA

Rodgers GE (1994) Introduction to coordination, solid State and descriptive inorganic chemistry. McGraw-Hill, Inc, New York

Samanta KA, Agarwal P (2009) Application of natural dyes on textiles.Indian J Fiber Textil Res 34:384–399

Scheyer LE, Chiweshe A (2001) Application and performance of disperse dyes on polylactic acid (PLA) fabric. University of Nebraska, Lincoln

Vickerstaff T (1954) The physical chemistry of dyeing. Oliver and Boyd, London, pp 1–110

Index

W. A. Ahmad et al., *Application of Bacterial Pigments as Colorant*,
SpringerBriefs in Molecular Science, DOI: 10.1007/978-3-642-24520-6,
© The Author(s) 2012